EQUATORIAL GUINEA

WESTVIEW PROFILES • NATIONS OF CONTEMPORARY AFRICA
Larry W. Bowman, Series Editor

Equatorial Guinea: Colonialism, State Terror, and the Search for Stability, Ibrahim K. Sundiata

Mali: A Search for Direction, Pascal James Imperato

Cameroon: Dependence and Independence, Mark W. DeLancey

The Central African Republic: The Continent's Hidden Heart, Thomas E. O'Toole

Botswana: Liberal Democracy and the Labor Reserve in Southern Africa, Jack Parson

†*Kenya: The Quest for Prosperity*, Norman N. Miller

†*Mozambique: From Colonialism to Revolution, 1900–1982*, Allen Isaacman and Barbara Isaacman

Senegal: An African Nation Between Islam and the West, Sheldon Gellar

†*Tanzania: An African Experiment*, Second Edition, Revised and Updated, Rodger Yeager

Ghana: Coping with Uncertainty, Deborah Pellow and Naomi Chazan

Somalia: Nation in Search of a State, David D. Laitin and Said S. Samatar

Ethiopia: Transition and Development in the Horn of Africa, Mulatu Wubneh and Yohannis Abate

Zambia: Between Two Worlds, Marcia M. Burdette

São Tomé and Príncipe: From Plantation Colony to Microstate, Tony Hodges and Malyn Newitt

†Available in hardcover and paperback.

EQUATORIAL GUINEA

Colonialism, State Terror, and the Search for Stability

Ibrahim K. Sundiata

Westview Press
BOULDER, SAN FRANCISCO, & OXFORD

DT
620.22
$.586$
1990
155.361
$ma.1992$

Westview Profiles/Nations of Contemporary Africa

Published in 1990 in the United States of America by Westview Press, Inc., 5500 Central Avenue, Boulder, Colorado 80301, and in the United Kingdom by Westview Press, Inc., 36 Lonsdale Road, Summertown, Oxford OX2 7EW

Library of Congress Cataloging-in-Publication Data
Sundiata, I. K.
  Equatorial Guinea : colonialism, state terror, and the search for stability / by Ibrahim K. Sundiata.
    p. cm. — (Profiles. Nations of contemporary Africa)
  Bibliography: p.
  ISBN 0-8133-0429-6
  1. Equatorial Guinea. I. Title. II. Series.
DT620.22.S86   1990
967'.18—dc19
                                                                                    87-14749
                                                                                         CIP
                                                                                         r89

Printed and bound in the United States of America

⬡   The paper used in this publication meets the requirements
(∞) of the American National Standard for Permanence of Paper
    for Printed Library Materials Z39.48-1984.

10   9   8   7   6   5   4   3   2   1

Contents

List of Tables and Illustrations   vii
Preface   ix

1 The Land and People   1

The Geographical Setting, 2
The People of Rio Muni, 9
The People of Bioko, 13
Notes, 15

2 History   17

African Trade and European Interests, 18
The Advent of Spanish Colonialism, 25
The Colonial System, 30
The Colonial Economy, 35
Migrant Labor, 43
The Socioeconomics of Development, 48
Notes, 52

3 Politics   55

Decolonization, 55
The Regime of Macias Nguema, 63
The Regime of Obiang Nguema, 74
Notes, 88

4 The Economy   91

Decline, 91
Rehabilitation: Finance and Aid, 98

Exports and Infrastructure, 106
An Overview, 115
Notes, 116

5   *Society and Culture*   119

Religion, 121
Education, 131
Language, 134
The Status of Women, 136
Health, 140
Notes, 145

6   *Conclusion*   149

Notes, 153

Selected Bibliography   155
Index   163

Tables and Illustrations

Tables

2.1   Export of cocoa, 1894–1972   42
4.1   Cocoa production, 1968–1988   109
4.2   Composition of exports, 1970–1987   111

Figures

1.1   Equatorial Guinea in relation to West Africa   3
1.2   Rio Muni, Bioko, and Annobón   4

Photographs

Street in Bata, the capital of Rio Muni   6

Pico de Basilé viewed from the port of Malabo, Bioko   8

Headquarters of the Cámara Agrícola, Bioko   31

Abandoned main house of a European *finca*, Bioko   37

Election scene at a polling station in Rio Muni, 1968   63

Portrait medallion of Macias Nguema   67

President Obiang Nguema, with visiting U.S. naval
   mission   78

Logging camp, Rio Muni   113

Fishing boats on the beach at Annobón   114

Parade scene in Malabo   120

Wedding procession on Annobón   122

Centro Cultural Hispano-Guineano, Bioko   135

Luba Hospital, Bioko   143

Preface

In early 1970 I became the first outside academician to visit independent Equatorial Guinea. At the time the country was headed by Francisco Macias Nguema, one of Africa's most capricious and tyrannical dictators. As thousands can attest, arbitrary arrests and disappearances were common. In my somewhat less dramatic case, I was placed under house arrest and my research confiscated.

Three years after my return to the United States, I publicly expressed the view that Macias Nguema would eventually be brought down by elements within his own terror apparatus. Fortunately for the citizens of Equatorial Guinea, this came to pass six years later.

By the time I returned to Equatorial Guinea in the late 1980s, much had changed. Many things gave hope for the future; many others provided a continuing cautionary lesson. Today, more than a decade after the overthrow of the country's first dictatorship, a plethora of problems—economic dependency, ethnic rivalry, lack of infrastructure—remains. As the country moves through the 1990s, it will be imperative that basic socioeconomic problems be addressed. The unfolding history of Equatorial Guinea demonstrates that the vagaries of political leadership cannot be separated from the material constraints and challenges of nation-building.

The opinions expressed here are entirely my own. Assistance and counsel has been obtained from numerous people of various political tendencies in Africa, the United States, and Spain. Their help has been invaluable. I thank all those in Equatorial Guinea who aided me in my research, some of it undertaken under the most arduous conditions. I greatly appreciate the help of the foreign personnel in Malabo who helped me in my research, especially Cord Jakobeit and Robert Klintgaard.

I am also deeply grateful to those scholars working outside Equatorial Guinea, especially Max Liniger-Goumaz and René Pélissier, who have corresponded with me on a subject that until recently was a closed book to Anglophone readers.

Ibrahim K. Sundiata

1

The Land and People

The political decolonization of Africa in the latter half of the twentieth century created a number of independent microstates. The Republic of Equatorial Guinea is one of the oldest such countries. Since its independence in 1968, it has been joined by a plethora of new polities—Guinea-Bissau, the Cape Verdes, São Tomé and Príncipe, the Comoros, the Seychelles—that face the unpleasant dilemma of being juridically independent but economically dependent. This is a problem for all African nations; in the case of microstates, the problem is most glaring and the solutions most elusive.

Equatorial Guinea, formerly Spanish Guinea, is a paradox. Its major components are the mainland province of Rio Muni and the island of Bioko (formerly Fernando Po). The two areas have highly different histories that included, during the colonial period, greatly divergent patterns of investment and European penetration. These differences have conditioned their interrelationship and the country's stability.

While Rio Muni remained a colonial backwater devoted mainly to subsistence farming and lumbering, Bioko was highly engaged in the plantation production of export crops. In 1960 exports totaled more than $33 million, the highest level of exports per capita on the continent ($135). The major island, together with the smaller island of Annobón, had an annual per capita income of between $250 and $280.[1] Additionally, Bioko had one of the highest literacy rates in black Africa. Under Spanish colonialism, certain products fetched a price higher than world market levels. As a result of large-scale labor migration from Nigeria, the production of high-quality cocoa, principally from Bioko, reached a total of about 35,000 metric tons in 1968. Coffee was the second largest export crop, with annual production totaling 20,000 metric tons in 1968. Bioko was not only an exporter of cocoa, coffee, and bananas, but it also possessed some processing industries, such as chocolate and soap factories and palm oil works. The future seemed bright and three petroleum companies were prospecting for offshore oil.

1

By the 1980s, however, Equatorial Guinea was one of the least developed countries in the world. The little republic's population is, according to a 1989 estimate, approximately 389,000 inhabitants. In 1985 per capita income was $250; this represented little or no increase over the colonial period. But by the same year, exports had dropped more than 27 percent to $24 million.[2] As the country enters the 1990s, industry is almost nonexistent. Enterprises linked to cash crop production have not regained pre-1968 production levels, and many are closed or useless because of lack of maintenance. A concomitant of this inactivity is that there are not more than 1,800 to 2,000 salaried workers.

The country could be and has been seen as a classic case of postcolonial decline. The first president, Francisco Macias Nguema, went a long way toward dismantling the colonial economy and liquidating those Guineanos linked to it. In the eleven years of his rule, 1968 through 1979, at least a third of the population was killed or went into exile. Socioeconomic decline is, however, more than just a result of dictatorial mismanagement. The nation's lack of a population sufficient to maintain export agriculture has been, and remains, an acute concern. In addition, Equatorial Guinea has been bequeathed illogical national boundaries and an export economy heavily dependent on cocoa. The postindependence collapse of arrangements meant to benefit Spanish plantation owners spelled the end of a manipulated prosperity. The trough into which Equatorial Guinea has fallen must be seen in relation to the artificial high arranged by Francisco Franco's Spain during the 1960s.

Under its second president, Teodoro Obiang Nguema, rehabilitation has proceeded slowly. One-party terror has been replaced by one-party authoritarianism. Under Obiang Nguema, the government has reversed the self-imposed isolation of the previous regime and assiduously courted an array of aid donors, including South Africa. It remains to be seen if, even with massive amounts of assistance, the country can overcome the handicaps imposed by underpopulation, small size, and a heritage of authoritarian rule. Both the international aid community and the government of Equatorial Guinea have pinned their hopes on surmounting these impediments.

THE GEOGRAPHICAL SETTING

For purposes of geographical discussion, the Republic of Equatorial Guinea can be divided into three parts: the continental province of Rio Muni, the island of Bioko, and the islets of Annobón, Corisco, and the Elobeys (see Figures 1.1 and 1.2).

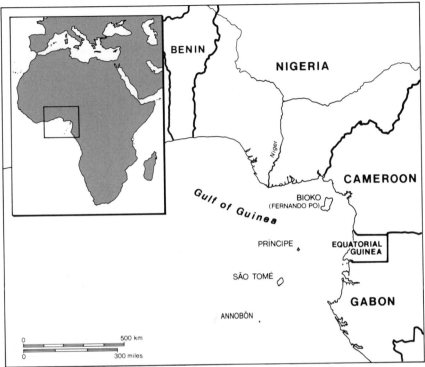

FIGURE 1.1 Equatorial Guinea in relation to West Africa. (Courtesy Seán Goddard)

Rio Muni

The largest portion of the republic is Rio Muni, which has an area of 26,000 square kilometers. It is bounded on the north by Cameroon and by Gabon on the east and south. The coastline, although generally regular, has several capes: Punta Mbonda (or Uvomi), Cape San Juan, and Punta Itale. Most of Rio Muni is a peneplain with an average altitude of 650 meters. Rivers are considerably affected by tides, but surf is largely absent and ships can be loaded at several coastal ports by lighters. The narrow coastal plain rises gradually to a range of hills from 1,000 to 1,200 meters in the east. The territory's geology consists of several different types of rock formations: gneiss, gabbroes, diorites, and granite. The coast is made up of tertiary clays in the northwest and of secondary slates between the Mbini River and the southwest coast. In the north, the shore has some small rocky formations occupied by oyster beds; in the south, the coast is characterized by ria.[3]

Rio Muni has untapped subsoil wealth, including cretaceous coastal deposits and petroleum. The territory also has other minerals, such as

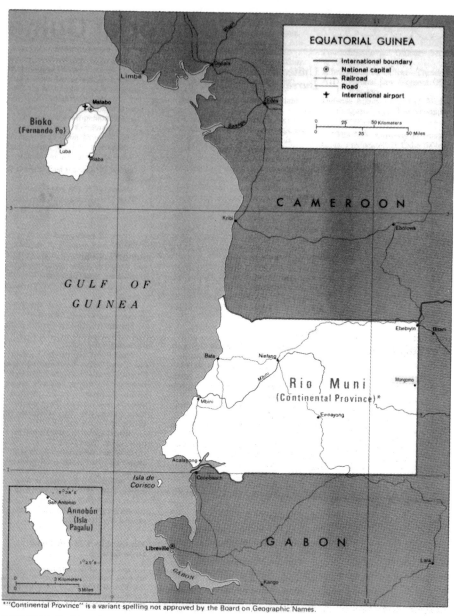

*"Continental Province" is a variant spelling not approved by the Board on Geographic Names.

FIGURE 1.2 Rio Muni, Bioko, and Annobón.

thorium, titanium, and gold. Coal beds in Cape San Juan continue south to Dambo (Dombo) and Gombie (N'Gambie) in Gabon. In the Montes de Cristal region, iron is found in association with sulfur. Iron also exists in the superficial strata of the shores of the Bañe River. There is quartz at Punta Cutia and in Cape San Juan there is excellent chalk and clay. There are also various types of sandstone. Surveys indicate the presence of uranium.

The mountainous areas in the east, the Montes de Cristal, are continuations of ranges in Gabon and Cameroon. The Montes de Cristal terminate in Monte Mitra, also called Monte de los Micos, or Biyemeyem. Another peak, Monte Chocolate, is in the center-west and is 1,100 meters in height; nearby Monte Alen is equally tall. To the north of the settlement of Evinayong is the 1,200-meter Monte Chime. Another outcropping, the Piedra de Nzas, extends to the north along the Rio Mbini. The coastal chain of the Paluviole Range extends from the southwest to the Seven Mountains in the northwest.

Rio Muni's average temperature is a warm 78° F (26° C), and the average humidity is around 86 percent. The climate is rainy; annual rainfall is 2,300 millimeters or more. The wettest times of the year are from February to June and from September to December—the reverse of the dry/wet cycle on Bioko. Even in the dry season the temperature runs from 78° to 82° F (26° to 27° C).

Most of the continental region is tropical rain forest. Overlooking the mass of vegetation are the trunks of gigantic trees, whose branches extend 30 to 50 meters. The largest of these species is the *ukola*, which rises to a height of up to 80 meters. More than 140 different plant species exist, including the okume tree, which is 35 to 45 meters in height and has a thorny trunk. Its wood—pinkish, tender, and light— is used in making plywood. Rio Muni shares the world's monopoly on this wood with southern Cameroon and Gabon.

In addition to the tropical rain forest that covers most of the territory, there are two other types of forest: the secondary forest and the river forest. The secondary forest is virgin rain forest that has been cleared of large trees and has had its underbrush burned in order to make way for farms. River forests follow watercourses and have their own type of vegetation. The roots of mangroves, which may reach a height of 5 or 8 meters, invade streams and rivers. Mangroves can form a wall on both margins of waterways even at some distance from the coast, especially in estuaries where numerous streams are subject to the influence of the tides.

The body of water that gives Rio Muni its name lies on a geological fault. It has sandbars at its mouth at Cocobeach, but they have not proved an insuperable barrier to its navigation. The "Muni River," in

Street in Bata, the capital of Rio Muni. (Photo courtesy Frank Ruddy)

fact, is a wide estuary with several rivers flowing into it: the Congüe, the Mandyani, the Mitong, the Mven, and the Mitémele (Utamboni). Kogo, an important trading center, is on the northern bank of the Rio Muni, 20 kilometers from its mouth. Upstream is Gande (or N'Gande) island, which once served as an important trade center.

Besides the rivers flowing into the Rio Muni estuary, there are several other important watercourses. Almost all of them flow from east to west. In the north is the Ntem (formerly the Rio Campo), which partially constitutes the frontier with Cameroon. It has a sandbar at its mouth but is navigable by small craft. Near Ebebiyin, in the far northeastern corner of Rio Muni, the River Kye, an affluent of the Ntem, enters Equatorial Guinea and then crosses into Gabon. The Rio Mbini (formerly the Rio Benito) forms the central axis of Rio Muni, cutting the territory in half, and originates some 850 meters above sea level. This river is frequently cut by rapids and waterfalls. The Mbini is navigable by small craft from the last rapids at Senye, 40 kilometers from its mouth. Larger boats cannot enter, however, because of a sandbar formed by estuarial tides.

Traditionally, watercourses have provided the routes for paths and, in turn, these paths have been connected by trails over the watersheds. Small rivers such as the Envia, Utonde, Eccu, Endote, Bicaba, Itembúe, and Ñaño have been used principally for the transport of lumber. Lighter wood, such as okume, is floated down these rivers as logs, whereas heavier wood is taken down by small craft.

The rivers of Rio Muni may be crossed in a number of ways. Wider rivers, such as the Ntem and Mbini, are crossed by dugout canoes. The Abanga, Ecucu, and Utonde, as well as other midsize rivers, are crossed by indigenously crafted suspension bridges. Smaller streams are traversed by bridges made of a single log, which sometimes have a guardrail made of vines. In swamps, vegetable-fiber mattings secured to posts driven into the ground are sometimes constructed.

The fauna is varied, although much depleted. In 1928 there were 138 different species; however, many of these animals have now disappeared due to overhunting. There are several significant mammals. Primates include macaques, mandrills, and chimpanzees. Gorillas were once well represented. Several types of felines also inhabit the region, including the wild, golden, and cerval cats. Rio Muni also has the guinea pig and the squirrel.

Reptiles and amphibians are well adapted to the ecology. Boas from 8 to 10 meters in length, plus the python and the African cobra, are also present. Tortoises of various sizes dwell in Rio Muni and its offshore islets; the most important are the sea turtle and the land turtle, or carey. The latter was especially abundant on the offshore islet of Corisco. There are two varieties of crocodile, *osteolaemus tetraspis* and *crocodilus calapharactus*.

The country has both European and tropical bird species. European species include sparrows, doves, ducks, swans, pheasants, cranes, eagles, herons, vultures, and falcons. Tropical species are also abundant and include parrots, hummingbirds, flamingos, ostriches, pelicans, and guinea fowl.

Insect life abounds. Present are various types of mosquito, including *anofeles*, *culex*, and *stegomya*, which are transmitters of malaria, filariasis, and yellow fever, respectively. Although the tse-tse fly, the transmitter of sleeping sickness, is present, there have been several near successful attempts to eradicate it. Chiggers, horseflies, driver ants, and termites also exist.

The sea off Rio Muni teems with life. Whales and sharks inhabit the waters off Annobón. Fish, including flounder, kingfishers, rays, sawfish, and giltheads, are an important economic resource. Other important marine life includes eels, squid, lobsters, and crabs.

Bioko

The island of Bioko, the site of the capital, Malabo (formerly Santa Isabel), has a total area of 2,017 square kilometers and is about 3° north

Pico de Basilé viewed from the port of Malabo, Bioko. (Photo courtesy Frank Ruddy)

latitude. At its longest, the island measures 75 kilometers; its average width is 35 kilometers. The island's history has been conditioned by proximity to the coast; at the closest point it is 32 kilometers from the mainland. Bioko is 259 kilometers from Bata, the capital of Rio Muni.

Bioko is extremely mountainous and geomorphologically resembles São Tomé, Príncipe, Annobón, and Corisco—all formed by extinct volcanoes: "[Bioko] was obviously once a peninsula of the Cameroon volcanic regions, and during some seismic disturbance the sea broke in between the [Bioko] mountains and those of the Cameroons." Further-more, "the islands and islets of Ambas Bay [off Cameroon] . . . are a remains of the former isthmus, which is further represented by a continuous ledge about thirty miles in breadth, still connecting [Bioko] with the mainland, under a shallow sea of from 200 to 290 feet [61 to 88 meters] in depth." On either side of the sunken isthmus "the depth of water suddenly increases to 6,000 feet [1,829 meters]; and between Bioko's southern extremity and the sister volcanic islands of Principe and São Thome the ocean depths are still greater (9,000 to 10,000 feet) [2,743 to 3,048 meters]."[4]

Bioko has three main peaks. The highest, the Pico de Basilé, is of volcanic origin and is an extinct crater. It rises to 3,008 meters and has a bottom some 152 meters below its summit. The southern portion of the island also contains the crater lake of Moka (previously Lake Loreto), which is some 1,524 meters above sea level. To the north of this lake, along the mountainous ridge of the island, are other lesser crater lakes. Small rivers and creeks are unusable because they flow only during the wet season. It is historically important that there are two excellent natural harbors: Malabo and Luba (formerly San Carlos).

Bioko's plant life is varied with 826 different species represented. The central region was once heavily forested. Cultivable lands lie around the perimeter in a band 7 kilometers wide and at an average height of 500 meters above sea level. Bioko's climate is tropical and the volcanic soil is ideal for growing cocoa, coffee, and tropical fruits. At the higher elevations the flora and fauna are of a temperate and largely European variety. The island possesses several geobotanical zones that are determined by elevation and prevailing winds. The uplands have proved resistant to the introduction of cocoa, the island's main crop. For instance, the elevated Ureka region of the south is swept by destructive violent winds.

Annobón, Corisco, and the Elobeys

Equatorial Guinea's small islands and islets have been of varying economic and historical importance. Annobón—formerly Palagú in the 1970s—is the only part of Equatorial Guinea south of the equator; it lies between 1° and 2° south. The island is 6 kilometers long and 3 kilometers wide and is located 670 kilometers from Malabo. Its rugged terrain has several peaks, the tallest over 600 meters high. Annobón has no good harbors or sheltered beaches. Palé (San Antonio de Palé) is the easiest point of entry and lies on the northeast coast. The population is largely concentrated in two small villages. There are two rainy seasons: April to July and October to November. There are only 117 plant species and the volcanic land is unsuitable for farming. Fishing has been the traditional mainstay of the islanders.

Off the coast of northern Gabon lie several islets that belong to Equatorial Guinea. The islands of Corisco, Elobey Grande, and Elobey Chico are found in the Rio Muni estuary. Corisco is 14 square kilometers in area. The Elobeys have a combined area of 2 square kilometers. They average about 10 meters above sea level, are flat, and lack fresh water.

THE PEOPLE OF RIO MUNI

The area that today is called Rio Muni and Gabon was settled by Bantu-speaking peoples in the second millennium B.C. Archaeological finds in the area are scanty, but numerous flaked tools shaped into hoes and axes have been found in the Gabon region. Some of the tools are partially polished and are found throughout the area.[5]

Before the arrival of Bantu speakers, the inhabitants were probably non-Bantu-speaking pygmies. Today these people do not exist, except on the islets in the Rio Ntem near the Cameroon border. European penetration caused their disappearance or isolation in part of the country. Subsequently, they mixed extensively with other groups. This has caused a shift in their somotype and a blending of culture between neighbors. For instance, the Fang, the predominant ethnic group, developed patron-client relationships with the remnant pygmy groups.

The migration of the Fang was preceded by the entrance of other Bantu-speakers, known generically as the Ndowe. During the colonial period the groups were called *playeros* (beach dwellers) by the Spanish. In the broadest sense the term refers to the Benga, Bapuku, Bujeba, Balenke, and Baseke.[6] The Ndowe are usually classified into two major groups: the Boumba, in the south, include the Benga, Bapuku, and Enviko; the Bongue, located in the north, include the Balenke, Baseke, Bomudi, Buiko, Asagon, Kombe, and other groups.[7]

The settlement patterns and interrelations of the various peoples are complex and changing, which has led to some confusion of ethnic and linguistic boundaries. Among the Bongue, the Balenke are the most eastern group. They live along the Mbini River and its affluents, as well as at the Muni estuary. The Baseke live between the Utonde and Ntem rivers. They seem to have an affinity with the Basa of Cameroon, but at present their language is, to a certain extent, mixed with Fang. According to one legend, the Bujeba had their origin near Ebolowa in southern Cameroon. They once occupied some twenty-five coastal villages to the north and south of Bata. The last group of the Bongue subdivision is composed of speakers of Kombe. They arrived after the Balenke and the Bujeba, perhaps spurred on by the influx of Fang. They are found in the Mbini, Ntem, and Bata regions. The Bongue language is also spoken by the Buiko on the northern border of Rio Muni. The Buiko are supposedly a mixture of a pygmoid population and Bantu speakers. The Boumba live to the north of Cape San Juan and now generally speak Benga. One of the Boumba groups, the Enviko (or Nviko, Mviko), populated the southwest coast along the lower Mitémele (Utamboni). Others, who were formerly enslaved by the Benga, live on Elobey Grande.

In general, the Ndowe migrated to the coast from a riverine region deep in the interior. The groups are farmers and fishermen and raise manioc, malanga, and bananas. According to some traditions, the peoples migrated to the coast and then away again as a reaction to the European slave trade. The Ndowe engaged in trade and conflict with the Mpongwe people of northern Gabon and were drawn into the European trade network. Contact with Europeans led to the spread of European diseases. By the nineteenth century the Ndowe were much reduced in numbers and had to face the growing power of Fang migrants from the interior.

The Benga were perhaps the last Ndowe group to arrive on the coast. They migrated after the Bujeba and the Balenke and occupied the region between Cape San Juan, Corisco, and Cape Esterias. At the turn of the nineteenth century they were on the southern shore of the Mbini River and later occupied the islands of Elobey Grande, Elobey Chico, and Corisco. Early in the twentieth century they still dominated the trade and politics of the Elobeys and Corisco.

Today approximately 80 percent of the population of Equatorial Guinea belongs to the Fang ethnic group. The Fang are part of a larger group generally referred to as Pahouin. The Fang are found in southern Cameroon and extend into Gabon as far south as the Ogowe River. Some traditions cite Azamboga (Azapmboga) in Cameroon as their point of origin. As early as 1817 an English visitor to the coast commented on the movement of the group.[8] Such travelers were especially intrigued by the tales of cannibalism told by the littoral peoples about the Fang. Migration continued into the last quarter of the century and probably followed two different paths. One group came from the Montes de Cristal escarpment in the 1840s and another from the upper Ntem region after 1860. By the late 1850s Mpongwe merchants in northern Gabon were seeking marriage alliances with Fang groups and bringing individual Fang to the coast for visits. Some groups continued to migrate south. By the early 1870s some clans were south of the estuary of the Ogowe River in Gabon. Around 1890 they reached their southernmost point. At the same time, the territory of the Fang itself was subject to raids by more northern Pahouin, particularly the Bañe.

Villages were extremely mobile; in one instance, a village moved more than 500 kilometers; another village moved more than 200 kilometers.[9] Migration was most likely motivated, at least in part, by a desire to eliminate African middlemen in trade with Europeans. Warfare with interior groups and mythico-religious beliefs also played a part. Movement was not directed by a centralized political authority but instead involved the gradual spread of groups engaging in shifting cultivation. Society was highly fragmented and egalitarian. Leadership within Fang groups depended on charisma and sometimes wealth. Because of the piecemeal nature of migration, there was great intermixture of groups, as well as the incorporation of remnants of peoples they found in their path. Kinship groups did maintain some ties, however. During the late nineteenth century, it was observed that in spite of this confusion [migration], the Fang nations [clans] have kept intact the union of the early days: the villages of one nation are separated often by enormous distances and by villages of a large number of different nations. Nevertheless, the connection is so real that the village can by law call upon the entire nation for support of its cause, just as the individual can call upon the village.[10]

According to a recent study, the Fang are divided into six major groupings, traditionally referred to as tribes: Ntumu, Okak, Fang-Fang (Fang proper), Mvae, Betsi, and Meke.[11] One of the main tribes in Rio Muni is the Ntumu. National boundaries have accentuated certain ethnic divisions. For example, the frontier between Rio Muni and Gabon increased the distinction between the Okak and Betsi-Meke Fang.

The same clans were distributed throughout the various tribes, and cultural, rather than genealogical, differences were the major delineators of tribal affiliation. Continental Equatorial Guinea contains about fifty clans. A Ntumu subgroup, the Esangui of the Mongomo district, has been particularly active in postindependence politics. With the decline of traditional village life in the twentieth century, the Fang have gone to considerable lengths to bind clans together and to create a pan-Fang unity (i.e., the Alor Ayong movement).

The Fang are patrilineal and patrilocal. They practice double exogamy and thus are forbidden to marry a person belonging to either the father or mother's clan. It is also forbidden to marry into a clan that had a common discernible origin with the mother's clan. Exogamy has the effect of creating a modicum of transethnic unity. At the same time, in the nineteenth century it was not undergirded by other transethnic associations. Although a number of concepts existed for groups beyond the village level, they were only vestigial and did "not function in either clan politics or in the religious life, such as they were."[12] The *ayong* tribe is a vague term with little sociopolitical reality. The extended family (*ndebot*) is of greatest importance in daily life.

The arrival of the Fang on the coast doubtlessly brought certain changes. The migrants evidently had little knowledge of navigation or canoe construction. They were agriculturalists and continued to grow their traditional crops, in addition to a number of cultigens known to peoples on the coast. In areas of cleared forest the Fang continue to grow yucca, yams, peanuts, malanga, and plantain. Traditional planting is seasonally rotated: Yucca is first, followed by peanuts, yams, and malanga. The Fang also have papaya, coconut palm, mango, and avocado. Some sugarcane for chewing and a small amount of tobacco are also cultivated. Palm oil is used in a sauce with foods. Other foodstuffs include "chocolate" derived from the "dika" tree, breadfruit, maize, and manioc.

In the early twentieth century, once every five to ten years, it was traditional for a village to shift its site when agricultural lands had been exhausted and dwellings had deteriorated. The agricultural cycle had well-defined stages. Toward the end of January, the Fang "go into the forest and find a favorable place, [and begin] the work of cutting . . . down the big trees. . . . In the beginning of March, the branches of trees are chopped up in pieces and burned, and soon, in the place of the great forest, there remain only the big trunks lying on the ground. In this ground, covered with ashes, the women place banana plants, stalks of manioc [cassava] and seeds of gourds."[13] In the next season of rains, crops ripened. After the harvest, the terrain was cleared again and replanted. In the longer of the two annual dry seasons the Fang fished in fallen river waters. Traditionally, fishing was succeeded by a period of hunting, until the onset of the next cultivation period.

In the nineteenth century, Fang weapons included crossbows, spears, elephant hide shields, and a few firearms. Europeans traded cheap manufactures and arms for skins, wood, and wild rubber. Ivory, rubber, and okume wood were major exports. The elephant (nsoc) lived in the interior and the Fang used traps in the form of pits to hunt them. A hunt was begun only after careful preparation; men of several villages usually participated. When they had surrounded an elephant or it had fallen into a trap, the animal was speared with a gigantic arrow made from the trunk of a tree and capped with an iron point. The coup de grace was administered by lances or flintlocks. Hunters used drugs to daze their prey. Traps were also used to capture leopards and panthers, which were rather indiscriminately called tigre by the Spanish.

The precolonial Fang used a currency, the bikwele (singular, ekwele), made of spearheads of various sizes.

THE PEOPLE OF BIOKO

By the fifteenth century, the island of Bioko was the only inhabited island in the Bight of Biafra. Recent evidence indicates that the indigenous population, the Bubi, had been long established. The "language cluster of Bioko" may have been one of the first language groups to break away from the mainland parent Western Bantu. Early Bantu-speakers themselves arrived in Gabon between 1500 and 1000 B.C. Recent scholarship views Bubi culture as generally reflective of that of Western Bantu-speakers at the beginning of their migrations. If this is true, early Western Bantu culture "was purely neolithic and the archaeological evidence must therefore relate to 'neolithic' sites, not sites from the Iron Age."[14]

If the sequence of cultures on Bioko is attributed to the Bubi, the island was settled by Bantu-speakers before the seventh century A.D. The four principal phases of stone-using cultures were the Carbonera, Bolaopi, Buela, and Balombe.[15] The Carbonera culture was produced by a people who lived principally on the northern beaches. Its creators used stones to pave their houses, produced palm oil, and had a wide range of clay vessels and finely cut stone implements. Palm nuts from Middle Carbonera have been radiocarbon dated to the late seventh century A.D.[16] The succeeding Bolaopi culture began in the eleventh century A.D. and belonged to a people whose early artifacts are frequently encountered in mounds near the sea. The fishing population associated with the Bolaopi culture made fragile pottery and very fine hand axes. This industry was, in turn, replaced by the Buela in the fourteenth century. Its remains are encountered throughout the island, although most frequently at an altitude of 500 meters above sea level and at the mouths of rivers. The final phase of indigenous stone culture on Bioko is the Balombe. The Balombe stone culture was studied by ethnologists in the early twentieth century. This culture was characterized by tool making, which arose out of the previous tradition and by very simple pottery.

The fact that the Bubi did not produce iron clearly separates them from the great majority of Africa's Bantu-speaking populations. In the nineteenth century material culture was sparse; its most obvious artifacts were digging sticks, canoes, conical straw hats, and short pieces of log used as pillows.

Oral accounts collected in the early twentieth century recall much more recent migrations than archaeological evidence indicates.[17] They probably reflect new Bantu migrants in the present millennium. The detailed migration traditions of the Bubi collected by Europeans may have been garbled in transmission or improperly set down. Nevertheless, they must be taken as more than just misshapen folklore. It is not unreasonable to suppose that Bioko received migrants from the mainland from time to time in the present millennium. In the 1820s it was observed that "[Bioko] has been peopled for the neighboring continent by malefactors and runaway slaves, who are determined to sell their liberty dearly, and any persons attempting to deprive them of it will have cause to regret their temerity."[18]

The Ureka area in the southeast appears, from several sources, to have received newcomers since the eighth century. It is hypothesized that in the last century "to the South of Fernando Po came such refugees from Principe (namely, descendents of Angola slaves . . .); these same refugees were already present by 1780 and since then have become completely Bubi-ized." Also the "Bubi themselves, however, distinguish today still between the 'Potugi,' (Portuguese) who as progeny of these refugees inhabit the districts of Ureka, Ariha, etc., and the genuine Bubi, although externally and in lifestyle the slightest difference is scarcely detectable anymore."[19] Early in the twentieth century an observer noted that "there is reason to suppose that occasional canoe-loads of Basa, Isubu, or Bakwiri people reached the east coast of Fernando Po [Bioko] at different times, and that the *few* resemblances in vocabulary between the indigenous Fernandian speech and the Isubu, Duala, and Basa may be partly due to the occasional arrival of colonists from the opposite coast."[20] The migration of non-Bubi Bantu-speakers to the island in fairly recent times cannot be ruled out and goes far to explain the southeast to north direction of many of the oral migration accounts.

The nineteenth-century islanders were matrilineal agriculturalists who practiced shifting yam cultivation. Another crop, the malanga, was cultivated only by women; harvesting began when the yam crop had been consumed. The indigenes also kept chickens, fished, and collected honey and palm oil. In preparing the latter, the nuts were plucked from the cones, placed in a pile covered with palm leaves, and left until they were almost putrified. They were then pounded in a mortar-shaped hole in the ground lined at its bottom with small stones. The pounding was done with a large stone or a wooden pestle. The inner kernels were taken out and thrown away. The remaining pulp was placed in a pot over a fire, after which Bubi women pressed the oil out with their hands. Palm kernels were not used on any great scale.

In the first half of the nineteenth century, political authority was diffuse; people were divided into many clan groups inhabiting particular districts. The subchiefs of the smaller political divisions claimed to be brothers of their respective overlords. Confederations were made up of smaller political units frequently in conflict with one another. The authority of political leaders appeared very limited. A chief's power was checked by village elders and age sets.

In the early nineteenth century, European contacts led to the development of trade linkages between the Bubi and the outside world. One result was some movement toward political centralization. The process was still underway when colonialism began to make its impression on the area.

NOTES

1. "Update," *Africa Report* (Washington, D.C., January 1969), p. 26. Also see *Spain in Equatorial Africa, Political Documents* 2 (Madrid: Spanish Information Service, 1964), p. 51.

2. See *The World Almanac and Book of Facts 1989* (New York: World Almanac, 1988), p. 672.

3. Max Liniger-Goumaz, *Connaître la Guinée Equatoriale* (Geneva: Editions des Peuples Noirs, 1986), p. 19.

4. Sir Harry Johnston, *George Grenfell and the Congo* (New York: D. Appleton and Company, 1910), vol. 2, p. 946. For a detailed discussion of the topography and economic resources of the republic, see Armin Kobel, "La République de Guinée Equatorialle, ses ressources potentielles et virtuelles. Possibilités de développement," Ph.D. diss., Université de Neuchatel, 1976.

5. P. de Maret, "Fernando Poo and Gabon," in *The Archaeology of Central Africa*, ed. F. Van Noten (Graz, Austria: Akademische Druck-u. Verlagsantalt, 1982), pp. 62–63.

6. Liniger-Goumaz, *Connaître*, p. 21.

7. Carlos González Echegaray, "Migraciones de los pueblos playeros de Rio Muni," in *Estudios Guineos* (Madrid: Consejo de Investigaciones, Instituto de Estudios Africanos, 1964), vol. 1, p. 22.

8. James W. Fernandez, *Bwiti: An Ethnography of the Religious Imagination in Africa* (Princeton, N.J.: Princeton University Press, 1982), p. 33.

9. Georges Balandier, *The Sociology of Black Africa: Social Dynamics in Central Africa* (London: Andre Deutsch, 1970), p. 132.

10. Christopher Chamberlin, "The Migration of the Fang into Central Gabon During the Nineteenth Century: A New Interpretation," *International Journal of African Historical Studies* 11, 3 (1978), pp. 429–456, citing A. Aymes, "La situation des populations untour de l'estuaire," Archives Nationales, Section d'Outre-Mer, Gabon, April 1872, vol. 3, p. 1.

11. Fernandez, *Bwiti*, p. 81.

12. *Ibid.*, p. 90.

13. Alan B. Mountjoy and Clifford Embleton, *Africa: A New Geographical Survey* (New York: Praeger, 1967), pp. 596–597, quoting L. Marton, "Le nomadisme des Fangs," *Revue de Geographie Annuelle* 3 (1909), pp. 497–524.

14. Jan Vansina, "Western Bantu Expansion," *Journal of African History* 25 (1985), p. 132.

15. Amador Martín del Molino, "Que sabemos actualmente del pasado de Fernando Póo," *La Guinea española* (March 15, 1962), p. 67.

16. P. de Maret, "Fernando Poo," p. 60.

17. Gunter Tessmann, *Die Bubi auf Fernando Poo: Völkerkundliche Einzelbeschreibung eines westafrikanischen Negerstammes* (Hagen: Filkwan, 1923), p. 12.

18. John Adams, *Remarks on the Country Extending from Cape Palmas to the River Congo* (London, 1823; reprint, London: Frank Cass and Co., 1966), p. 148.

19. Johnston, *Grenfell*, vol. 2, p. 955, citing Oscar Baumann, *Eine afrikanische Tropen-Insel: Fernando Poo und die Bube* (Vienna: E. Hölzel, 1888).

20. Johnston, *Grenfell*, vol. 2, p. 955.

2

History

European interest was first directed at the islands of the Bight of Biafra rather than at the mainland. Bioko and its neighbors, São Tomé and Príncipe, were first visited by the Portuguese in the late fifteenth century. The proximity of the islands has led to a number of misconceptions. In 1978 a Unesco conference proposed that Bioko be studied as an example of the growth of a sugar plantation society.[1] More recently, a writer on African affairs stated that the human rights abuses in Equatorial Guinea "are excrescences of Portuguese colonialism."[2]

Conclusions about the island portion of Equatorial Guinea based on the experience of its Lusophone neighbors are incorrect. The Portuguese never conquered Bioko and the island never had a sugar plantation economy. In 1507 Luis Ramos de Esquivel did attempt to establish a plantation on the eastern coast but failed. In 1772 the governor of São Tomé and Príncipe sent his son, Manuel Gomes da Silva, to reconnoiter the island. Although the establishment of a base was considered, nothing came of it. For their part, the Bubi were largely unaffected by, and uninterested in, the slave trade. Vessels of various European nations may have kidnapped individuals, but a regularized trade did not develop. Spain claimed Bioko in 1778, but it was not until 1910 that Spanish sovereignty was recognized throughout the island.

There are reasons for the difference between the largest Biafran island and its neighbors. Bioko lacked much easily cultivable land because of its topography. Given a similar mortality rate for Bioko and São Tomé, the smaller would have the advantage of being easier to control with a limited number of Europeans. Unlike Bioko, São Tomé and Príncipe are well placed for visits by sailing vessels. In addition, the lack of an indigenous population on the Portuguese islands meant that initial alien settlement was not met by resistance. Also, fifteenth- and early sixteenth-century firearms may have been of little use in the rainy climate.

17

AFRICAN TRADE AND EUROPEAN INTERESTS

In contrast to Bioko, Equatorial Guinea's smaller islands figure inordinately in early European contacts. Annobón, for example, became an entrepôt and agricultural center. The uninhabited island was visited by the Portuguese in 1471–1472. Jorge de Melo, the first person to hold rights to the island, sold them to Luis de Almeida, who brought in slaves from São Tomé. In 1592 the Portuguese established a subgovernorship dependent on São Tomé.

Portuguese rule was desultory. At times the inhabitants were semi-independent and, at others, subject to spasms of metropolitan control. In 1613 and 1623 Dutch expeditions visited and commented on the island's admirable prospects. Their slavers, like those of the Portuguese, used Annobón for provisioning. At the turn of the eighteenth century, the island supplied slavers bound for Angola with fruits, cattle, oranges, and coconuts.

The population was missionized by Spanish and Italian Capuchins in 1645, 1647, and 1654. In 1757 a priest was sent to minister to the islanders, but he abandoned his post because he feared an insurrection. Soon after, in 1770, two black Catholic missionaries were sent. After almost a year, they also left, in this case because of the cultural resistance of the Annobonese, who had a strong sense of their own identity; the islanders had already evolved their own language, Fa d'Amo, an Afro-Portuguese creole.

The history of the little island of Corisco parallels that of Annobón. Portuguese slavers were especially active in the sixteenth and early seventeenth centuries. The Dutch occupied the island in the middle of the seventeenth century but soon left. In 1656 the Portuguese established the Companhia do Corisco to concentrate on slaving in the region. In 1723 these operations were transferred to Cape Lopez in present-day Gabon.

In the late eighteenth century, Portugal's Iberian neighbor hoped to establish its own slaving base in the region. In 1777 Spain purchased Annobón, Bioko, and Corisco. Given the success of São Tomé and Príncipe, Spain thought to use Bioko as a entrepôt for direct African slaving. Traditionally, slaving contracts (*asientos*) had been granted to non-Spanish companies. The desire for an African base was an early indication of the channelization of Spanish capital into the trade, a tendency that became more marked in the next century.

The Spanish were unaware of the ecological differences between the new acquisition and São Tomé and Príncipe. In 1777 they agreed to effect the transfer. A subsequent treaty, the Treaty of the Pardo, sealed the cession. In April 1778, an expedition left Montevideo, Uruguay, with

150 men and provisions for more than one year. The expedition arrived on Príncipe on June 29 and awaited Portuguese officials. Finally, in October 1778 Spanish and Portuguese envoys set foot on Bioko and Spain claimed possession.

What should have been the beginning of a continuous Spanish presence in sub-Saharan Africa soon went awry. In November the expedition sailed on to take possession of Annobón after brief stops at São Tomé and Principe. En route the leader of the flotilla, the Conde de Argelejos, died and was replaced by Lieutenant Colonel Joaquín Primo de Rivera. The latter was unable to claim Annobón because of the hostility of the inhabitants.[3] Rebuffed, he sailed back to São Tomé. Late in 1779 his expedition returned to Bioko and landed on the eastern side of the island at a bay christened "Concepción."

The Arrival of the Slave Trade

A member of the Spanish expedition, José Varela y Ulloa, surveyed the possibilities of successful slaving. He noted that Spain could not soon expect a tremendous influx of labor from the region. In spite of its former reputation, he was unimpressed with Annobón: "Annobón is a small island with acidic, dry soil in many parts. It has no cover or any type of natural protection that would afford a year-round anchorage site." The winds were unfavorable for sailing ships and the island was too far from the coast. Bioko was better situated "because if, at some time in the future, abundance of food and goods does reign on Fernando Poo [Bioko], then it is very possible that the [slave] traffickers of Juda (Whydah) will come to obtain provisions, water, and firewood." The trade south of the Cameroon River was not particularly brisk, but "along the Campo [Ntem] and San Benito Rivers [Mbini] and in the Cape of San Juan region, the Negroes are very brave and aggressive. . . . They usually come on board with ivory, wax, and dyed sticks [dyewood]." On the whole, the expeditionary was not very sanguine. He recommended that "in order to make the venture solvent, the only thing I can suggest is to build an establishment on the Gabon River and another on the Bay of Lopez Gonzalvo [Cape Lopez]. . . . If our sights do not go beyond the rights that the Portuguese have ceded to us, we cannot count on obtaining even 150 slaves each year." Most tellingly, he informed his superiors that the areas intended for Spanish slaving were not under Portuguese control and were probably worthless. Simply put, "the Portuguese do not have any rights to this island [Bioko] save that of discovery, because they have never established themselves on it; nor have they ever conducted any commerce with its inhabitants."[4]

As later expeditions were to discover, the climate proved unhealthy to Varela and his comrades. The morale of the settlement fared badly as mortality soared and in September 1780 mutiny erupted. Two days after the mutiny the entire expedition returned to São Tomé. Finally, in disillusionment, the small flotilla left for U.S. waters in December 1781.

Eighteenth-century attempts to resuscitate government-backed slaving on the lower Guinea coast met with no success. In 1781 Bernardo de Yriarte, a member of the Council of the Indies, proposed an all-Spanish slave trade under government supervision. In 1785 and 1786, José de Galvez, marqués de Sonora, contemplated another colony on Bioko. Spanish investors in the Caracas Company organized the Philippines Company with the marqués's patronage and were given an exclusive *asiento* for slave trading with the newly acquired island. Full-scale plans were cut short by Galvez's death; the initiative was short-lived and the company never supplied any slaves.

In spite of the collapse of strong support for government-sponsored activity, Spanish traders' interest in the Biafran area continued. After the abortive Argelejos mission, Spaniards continued to value the area for slaving purposes. From 1785 to 1800 many Spanish ships sailed from Corisco to the Caribbean. Unfortunately for slaving interests, the disruptions of the Napoleonic Wars militated against the maintenance of permanent factories.

In the period of British antislaving activity after 1807, Corisco and the surrounding area increased in importance. In an 1817 Anglo-Spanish treaty, Spain agreed to suppress the trade north of the equator immediately and to abolish all Spanish slaving in 1820. British cruisers were given a limited right to search and detain Spanish slavers. The treaty was signed at the very time that the Spanish slave trade began to burgeon. A stronger treaty in 1835 failed to stop the outflow of slave labor to the Caribbean. In the 1830s the Corisco region was visited by more than 100 Spanish ships per year. In November 1840 the British took decisive action. They attacked the island and captured *Miguel Pons*, one of the largest Spanish traders.

Even after the attack, however, the island retained its importance as a point of entry to central Africa. In the 1850s the traveler Paul Du Chaillu wrote:

> Though but a small island, Corisco has its hills and valleys, forests and prairies, and has even a little lake. . . . [There are] little native villages. . . , with their plantations of plantain, manioc, peanuts, and corn showing through the palm groves.
> Its population, of about 100 souls, is scattered all over the island. They are a quiet, peaceable people, hospitable to strangers and fond of white men. . . . They belong to the Mbenga [Benga] tribe, who are the most

enterprising traders and the most daring boatmen of the coast. They were formerly the most warlike tribe of this part of the country. . . . They are no longer so quarrelsome, and have lost that reputation for ferocity which formerly they prided themselves on.

This tribe inhabits not only Corisco, but also the land about the neighboring Cape Steiras [Esteiras] and St. John [San Juan].[5]

The Benga served as middlemen in the slave and other trades. One unfortunate consequence of European visits was a high rate of venereal disease. Corisco was famous for the trading acumen of both its men and women. The "isle of love" had a large number of "mammies," who provided trade goods and refreshment for European sailors. Unfortunately, as their population stagnated on the coast, the Benga were unable to push back migrants from the interior. They were also unable to stop disunity within their own society. After the death of Bonkoro I in 1846, the succession was disputed by his son, Bonkoro II, and several important leaders, most notably Munga (1848–1858). Munga won the dispute on Corisco and Bonkoro II fled to the mainland and established his capital at Cape San Juan. His political authority was aided by the movement of people from Corisco to the coast, as well as the addition of people from Cape Esteiras. Many Benga chiefs did not recognize the authority of either Bonkoro II or Munga. Bonkoro II died in 1874 and was succeeded by his brother, Manuel Bonkoro, who was a client of the Spanish. On Corisco, Combenyamango succeeded Munga and ruled from 1859 to 1883. In addition to succession disputes, the Benga found their trade was increasingly subject to scrutiny by the British.

Antislaving and Palm Oil

Significantly, whereas Corisco was seen by many Spaniards as the gateway to slaving, Bioko was seen by many British as the base for its destruction. As early as the mid-1820s, the British government had become convinced that the larger island was the logical base for patrols in West African waters. At least 15,000 slaves were being exported annually from Bonny and Old Calabar, the two main slaving ports. The state of medical knowledge pointed to Bioko as a "natural hospital" for invalids from the low and marshy shores of the mainland. In 1827 Captain William Owen established Clarence (now Malabo) and began landing recaptured slaves, mostly Igbo. In 1829 he was replaced by Colonel Edward Nicolls, who hoped to transform the island into a prosperous plantation economy using free labor. The base functioned until 1835.[6]

When the decision to abandon Bioko was made, protest was voiced by British merchants. In 1833 twenty-four firms warned that the island would become a center of Spanish slaving. More important, they pointed out that the British importation of palm oil had greatly increased since the occupation of 1827. Some hope remained that commerce and antislaving might still be combined. In the summer of 1837 Thomas Fowell Buxton, one of the leading parliamentary abolitionists, conceived of a scheme to open the heart of Africa by navigating the Niger. As part of the plan, Bioko would be purchased from Spain. Under a system of free trade, it would prosper as Singapore had under Sir T. Stamford Raffles. By 1838 the British government was convinced that the island was essential for its mission in West Africa. From 1839 to 1841 the British attempted to buy the island, but Spain refused. The ill-fated Niger Expedition of 1841 used nominally Spanish Bioko as its base. The failure of the expedition, due to high mortality, ended official British interest in the island.

Throughout the 1840s and 1850s British interests remained the only alien ones on the island. Private company rule appeared to succeed government rule when the government installations were sold to Richard Dillon and Company of London. Dillon hoped to break Liverpool merchants' hold on the Niger Delta palm oil trade. The company went bankrupt through mismanagement in 1836. The West African Company, which bought out Dillon, was composed largely of his London creditors. The company used Bioko as a transshipment point for gum copal, gum senegal, coffee, and grain. For most of its brief existence the company's chief exports were palm oil and timber. Company attempts to make the recaptured slaves into landless laborers were largely unsuccessful. The British government did not support the company's position in what was nominally Spanish territory. The firm's problems were compounded by the encroachment of Liverpool traders. Coercive methods and attempts at trade monopolies did little to increase profits. In 1843 the company's last shipment to England was only 98 casks of palm oil.

In late 1842 a British naval officer suggested that the British government buy out the firm. In 1842 the African Agricultural Association hoped to raise £40,000 and to encourage nonslave agriculture through black immigration from the United States and the Caribbean. The scheme failed for lack of capital. The British Baptist Missionary Society, whose missionaries had arrived on the island in 1841, acquired the West African Company's properties in 1843. The company had decided to abandon the exploitation of the island after suffering losses amounting to £50,000.

The Baptists brought settlers from Jamaica and ministered among recaptured slaves. The latter, in addition to immigrants from Sierra Leone, became the nucleus of a westernized black trading population, which came to be known as "Fernandino." They gained their livelihood through middlemen trading in palm oil. Although Bioko was not a major center of production, it was an important transshipment point for oil coming from the Niger Delta. By the early 1840s, the trading community had grown to over 800.

From 1835 until 1858 the little settlement was under no more than nominal European control. An Englishman, John Beecroft, was the acknowledged leader. His position was ratified in 1843 when Madrid recognized him as governor. In 1849 he was also appointed British consul for the Bights of Benin and Biafra, an area that covered not only the Niger Delta and Lagos, but Dahomey as well. Beecroft's deputy was William Lynslager, another European trader. After Beecroft's death in 1854, Lynslager served as governor until he was replaced by a Spaniard in 1858.

Alien settlement fostered trade, which had repercussions on Bubi society. In response to new opportunities for trade, its diffuse political organization underwent some centralization. The emergence of a paramount chief, Moka, probably reflected the changing economic order. Inland villages paid duties to those on the coast. Interestingly, this did not give rise to a group of Bubi merchants; the island's people concentrated on production rather than trade. They also consciously endeavored to minimize the impact of outsiders. Although Moka scorned the use of European goods, he did derive some benefit from trade. Footpaths led to his residence from all directions and, at times, he forced his people to detour in order to pay tolls.

The freedmen of Clarence (Malabo) were the middlemen between the Bubi and traders like Beecroft and his conferees. In the 1840s, Beecroft set a pattern followed by his successors. He and his associates owned shops that supplied the petty traders with European manufactures. The merchant-administrator had a special relationship with Robert Jamieson of Liverpool and used it to best advantage. He had his own trading settlement on the western side of the island, which gave him an early alternative port to Malabo.

Some petty traders worked directly for Beecroft and used his trade goods and his trading post. Others were tied to him more informally. The western side of Bioko was not only the location of Beecroft's establishment but was also a major trade center for numerous small-scale merchants. The most important area was on the opposite coast, particularly near Riaba (Concepción Bay). The environs contained at least 7 markets and in 1845 approximately 200 people traded there.

In the 1840s the profits of the palm oil trade ran from 150 to 300 percent. The island exported 300 tons of oil per year in the 1850s and 400 tons per year in the 1860s. Palm oil continued to be exchanged for goods and, by the 1870s, for cash. With its scanty population, Bioko was increasingly unimportant to the trade as a whole, however. In 1855 the island produced 360 metric tons of palm oil compared to the 16,383 metric tons produced in New Calabar and Bonny, 2,317 metric tons produced in Brass, 4,156 metric tons produced in Old Calabar, and 2,144 metric tons produced in Cameroon.[7] In the late 1870s only about 500 puncheons of oil were shipped per year at an average of 568 liters each. The value of this export was estimated to be $30,000.[8]

The Rise of Cocoa Farming

In the late nineteenth century a new crop, cocoa, was increasingly important. Cocoa was brought to São Tomé from Brazil in 1822. São Tomé pods were first taken to Bioko fourteen years later. From here the crop may have diffused to other parts of West Africa, such as Ghana. In 1879 the amount exported was approximately 100,000 pounds [45,359 kg], valued at about $20,000. The failure of the Spanish capital to thrive in the colony left room for the entrance of small-scale black cocoa farmers. Men who had acquired a stock of capital in palm oil trading invested it in cocoa as the prospect for greater gain was perceived. Bioko was a vent for ambition that was sometimes stifled elsewhere. For example, the most successful black planter of the nineteenth century was William Allen Vivour, a Sierra Leonean immigrant and trader. By the end of the 1880s, a few agriculturalists had begun to trade directly with British trading houses, among them Vivour, whose cocoa plantation at San Carlos (Luba) supposedly was one of the best in West Africa. By the mid-1880s he was the largest landowner on the island and employed more than a hundred workers. After his death in the 1890s, his widow, Amelia Barleycorn Vivour, owned the largest cocoa plantation on the island—400 hectares in Luba.

Agriculturalists faced a shortage of labor and, as a result, the exploitation of cocoa was slowing. By the end of the nineteenth century, only 6,500 hectares had been conceded. On Bioko, ties between employer and employee were largely transitory, involving employment of strangers for a specifically contracted period. Although there was some movement of migrant labor into farm ownership, major emphasis was on the development of plantations (*fincas*) dependent on an alienated and transient labor force. West Africa may be the realm of the peasant producer, but Bioko stands out sharply as the area of the Hispano-African *fingueros*.

Cocoa farmers had to look far afield for labor. In the 1880s Vivour employed an ethnically mixed force, in which the majority of the workers were from the Loango Coast. In addition, he employed some 30 workers from Accra, who performed basically artisan jobs, a few Grebo and Bassa from Liberia, and 4 Bubi. Others obtained labor from Batanga or

Bimbia in Cameroon. With the passage of time, there was a diminution in the number of laborers from equatorial Africa and in increase in the numbers from Nigeria, the Gold Coast, and Sierra Leone. Around 1896, African labor recruiters were obtaining labor from the area around Lagos and Ijebu-Ode in western Nigeria.

THE ADVENT OF SPANISH COLONIALISM

Nineteenth-century Spanish activity in what became Equatorial Guinea was a sporadic affair. Spanish slavers did an extensive business, but the Spanish government exerted no influence. Madrid became concerned that such neglect would disallow its claims, and by 1840 this again seemed imminent. In 1841 the British burned the slaving base of Pedro Blanco in the Gallinas region south of Sierra Leone. Having already burned bases on Corisco, they threatened to do the same thing to entrepôts run by Spaniards at Cape Lopez. Further concern was prompted by construction of a French antislaving base in Gabon.

Concern moved Madrid to act, and in 1843, it sent out an expedition under Captain Juan José de Lerena. After visiting Bioko and giving Clarence the name Santa Isabel, the expedition went on to the African mainland. There the Spaniard reportedly secured the allegiance of the Benga. Later Spanish claims rested on the assertion that he had also received the submission of the leaders of the Kombe, Bapuku, Enviko, and Balenke peoples. A brief 1846 visit by another official, Adolfo Guillimard de Aragon, served mainly to reassert an official Spanish presence.

In the wake of this visit, Madrid's cultural and political influence remained almost nonexistent. British missionaries and commerce seemed to claim the territory. The seizure of Spanish vessels suspected of slaving continued to create loud protests and demands for action. A furor was created in 1854 when the appointment of Domingo Mustrich as governor of Bioko was blocked by the British, who accused him of slaving. In 1858 the Sociedad Económica de Barcelona petitioned the government to encourage rapid colonization. Companies like Vidal and Ribas and de Montagut and Company also pushed for a Spanish commercial presence. These sentiments were echoed by Madrid's Sociedad Económica Matritense.

Spain experienced a spasm of imperialism in the mid-nineteenth century; colonial adventures were undertaken in Morocco, Indochina, Santo Domingo, and Peru. In keeping with the colonial upsurge, the government decided to occupy Bioko and parts of the coast. In 1858, eighty years after claiming the island, Madrid sent the first Spanish-born governor, Captain Carlos Chacón. The English Baptist missionaries were expelled to Victoria (Limbe) in Cameroon and Spanish Jesuits introduced. There was some hope that colonization and the production of tropical produce might create a *Cuba Africana*. Chacón brought a prefabricated hospital, medicines, and supplies for six months. The new governor almost immediately began plans for construction of roads, an improved dock, and a harbor light. An administrative apparatus was installed and a governor's council created.

In 1859, in spite of the record of Bioko's past insalubrity, European colonization was begun. Circulars were sent to the Spanish provinces encouraging emigration and promising free passage. A new governor, José de la Gándara, arrived with 128 colonists and 166 military personnel. In 1860 some colonists went up to the higher elevations, initially with favorable results. In spite of this promise of a "white highland," the Europeans soon died in great numbers. In 1861 the colonists were removed to Corisco, where conditions improved very little; half died. Yellow fever was brought from Havana in March 1862, and in two months, 78 out of 250 whites died. Bioko did not become another Canaries or, at the minimum, another São Tomé. Mid-nineteenth-century Spanish settlers met the same epidemiological barriers as their eighteenth-century predecessors.

The attempt at free Iberian colonization was a debacle. In the aftermath, Cuban freedmen (*emancipados*) were put forth as the answer to the problem of colonizing tropical Africa. In the early 1860s, a stream of royal decrees called for the use of black Cubans. One decree proposed enlisting and sending 75 or 80 emancipated Cubans to replace half of a white infantry company sent in 1859. It was thought that, besides serving as soldiers, they could be artisans and laborers. The emigration scheme came to fruition in 1862. Two hundred black emigrants, indentured for seven years, arrived from Havana. The government was encouraged and ordered another 200 emigrants. Emigration had the support of the home government, but it soon became evident that there was a lack of coordination with the Cuban authorities. Requests for further emigrants met with no response. Although *emancipado* emigration received a further blow in September 1866 when it was prohibited to transport freedmen overseas, the idea remained. In 1872 when the Ley Moret, which partially abolished Cuban slavery, went into effect, it gave the freedmen the choice of remaining in Cuba or going to Spanish Africa.

The cessation of the emigration of freed blacks did not mean the cessation of Cuban immigration. Penal settlement held out certain attractions. The dilatory attitude of the Cuban authorities on black emigration was matched by the alacrity with which they encouraged the expatriation of political dissidents. Such colonization had begun as early as 1861 when a *presidio* had been created and 13 prisoners transported from Malaga. In 1866, 19 political prisoners arrived as a consequence of a republican and socialist movement in Andalucia. Over 100 political deportees also arrived from Havana in 1866. Two hundred and fifty Cuban rebels arrived in May 1869. The penal scheme for Bioko ended the same year. Royal orders for the abandonment of deportations had already been issued and the repatriation of the Cuban political prisoners was begun at the end of the 1860s. Penal settlement, from Cuba or from the metropole, continued to be proposed throughout the last quarter of the nineteenth century. In 1875 the Real Academia de Ciencias Morales y Políticas debated the question. One cogent argument against penal settlement was that it amounted to a death sentence.

After plans for large-scale colonization in the 1860s were abandoned, there was a retrenchment in colonial policy. After a revolution in 1868, the new Spanish government decided to review its presence in sub-Saharan Africa. The findings of the governmental commission are highly illustrative of the colonial malaise:

> From 1858 to the present 50 million *reales* have been spent and, in spite of this sacrifice, there is not one meter of road, nor one solid bridge, nor even one masonry building, nor one newly created town, nor one native or Bubi conquered for Spanish civilization, everything remaining as it was twelve years ago. It is also true that the two expeditions of colonists sent at the expense of the Government have returned almost in their entirety, some for lack of aid . . . and others because of endemic illness. . . . And if it is certain that agriculture moved forward somewhat and that at first commercial enterprise was fomented, nevertheless, it was without rooting colonization there. . . . It is urged, then, to determine if that country contains enough favorable conditions for the creation by the State of an advantageous Spanish province . . . or if it will be more convenient to absorb the expense and abandon this project.[9]

The sub-Saharan territories were retained. A colonial decree fused the offices of governor and chief of the naval station in an attempt at economy. A small administrative staff was maintained on Bioko. Looking at the situation, one official commented: "How could we have competed with the English influence on Fernando Po [Bioko], if far from resisting it in our Spain, it is dominating us more and more every day. . . . To battle against England on any matter in Spain would be grand and patriotic, but to battle her here [in Africa] is pitifully ridiculous."[10] Mainland Equatorial Guinea remained largely untouched by Spanish colonialism and the Biafran islands claimed by Spain were largely ignored.

Between 1868 and 1900 Spanish Guinea was dominated by foreign capital and figured very little in metropolitan calculations. In the mid-1880s English shipping arrived four times per month and often stopped at San Carlos (Luba) and Concepción (Riabba) bays to take the island's produce to Great Britain. The German Woermann Line arrived, at most, twice monthly. Except for naval vessels, the island was seldom visited by Spanish ships. Money from the metropole arrived irregularly and the government was almost always in debt. English money was the medium of exchange, although old Spanish pesetas, worthless in the metropole, were in circulation.

On Bioko, the Bubi were barely aware that the Spanish claimed sovereignty. In the 1860s the Spanish met with scant success in attempts to win either the islanders' allegiance or labor. In 1859 the newly installed Spanish governor interviewed some chiefs and the next year reviewed some military formations. Neither this governor nor his successors were able to exert their authority throughout the island. For the most part, the attempt was not even made. In the 1860s what Spanish contact there was came through a Jesuit mission. Although the Jesuits had some success in the countryside, the curtailment of their activities after 1868 prevented them from gaining a firm following. It was not until missions were opened by the Claretian order in the 1880s that Catholic missionary activity received new impetus. In 1884 the missionaries made contact with Moka. Three years later they unsuccessfully attempted to get him to provide labor and acknowledge Spanish rule. Moreover, the governor of the colony visited Moka thirteen years later. The Bubi chief favorably impressed his guests by flying the Spanish flag and the governor attended a meeting of Moka's council. Nevertheless, labor and taxes were not provided to the European government.

On the other islands and on the mainland, Spain's position was even weaker. During the "Scramble for Africa" a German warship visited and almost claimed Annobón. It was only deterred when a Spanish priest hastily ran up his country's flag. A few interests championed sub-Saharan colonization. In 1881 Elias Zerolo, the director of the *Revista de Canarias* noted that "Spain possesses there [in the Bights], besides the islands of Fernando Po [Bioko], Annobón, Corisco, Elobey Grande and Elobey Pequeno, the immediate coast of no small area. For certain, the governments that have succeeded in Spain since 1858 have looked with the greatest indifference [on] the possession of that coast."[11] In 1883 the Congreso de Geografía Colonial y Mercantil called for the stimulation of trade, immigration, and agriculture. It also advised a subsidy for a steamship line. In addition, in order to protect mainland claims, the group urged the Ministry of Overseas Territories to support plans for central African exploration.

Spain's foremost tropical traveler, Manuel Iradier y Bulfy (1854–1911), made two trips to Africa, but had trouble persuading his compatriots to occupy the territories claimed. In 1875 he visited Bioko and Corisco.

He also visited Rio Muni and never got more than 100 miles from the coast. In 1877 Iradier, who was deeply inspired by the exploits of Henry Stanley, drew up an ambitious plan of exploration that included the interlacustrine region. Funding was not available, however, and it was not until 1884 that he returned to Africa.

Iradier's mission and that of his countryman, Amado Ossorio y Zayala, were intended to secure a large central African empire. Ossorio concluded a purported 379 treaties with indigenous leaders. At the Berlin Conference of 1884–1885, Spain claimed an area of 180,000 square kilometers in Africa. Unfortunately for Madrid, Britain claimed the territories in what became Eastern Nigeria. The German annexation of Cameroon in 1884 effectively frustrated Madrid's idea of acquiring the coast nearest Bioko. In the area south of the Cameroonian boundary, Spanish interests collided with those of France. Denmark was asked to mediate the boundary questions in 1892, but the issue was not settled until 1900. Then, in the wake of the Spanish-American war, Spain was left with a tiny enclave almost surrounded by French Gabon.

Around the turn of the twentieth century rubber and ivory were Rio Muni's chief products. Each of the various European trading posts employed between 30 and 40 Africans. The majority of Africans employed by Europeans were Benga or Kru and Bassa from Liberia. English firms (John Holt and Company, Hattan and Cookson, Forster) and German firms (Woermann, Kuderling, Randa-Skind, Schulze, Lübke) were firmly established. Some Spanish companies had also established themselves: Vidal y Rivas, Montagut y Compañia, Antonio Cuca, Antonio Trillos, Jaime y Miquel Eu. In 1887 the Compañia Transátlantica entered the area under the direction of Emilio Bonelli.

In addition to foreign traders, Rio Muni also contained a non-Spanish missionary presence. American Presbyterian missionaries had long been established on Corisco. In 1850 Rev. J. L. Mackey began work there. Ten years later the mission had spread its influence to Cape San Juan, Mbini, and Bata. In spite of its longevity, the mission did not gain great numbers of converts. After forty years of existence it had only 1,090 members in 9 centers. One-hundred and nineteen of these were in Bata and 230 were Mbini (Rio Benito). Most of the rest resided in territory that came to be French. As the Spanish claim to Rio Muni grew, the U.S. mission faced increasing opposition from the authorities. In 1875 the church on Corisco was given an African pastor, J. Ikege Ibia. The pastor, who had been trained in the United States, was subsequently exiled to Bioko at the request of Spanish Catholic missionaries. Similar tensions existed on Bioko, where English Primitive Methodist missionaries had begun their operations in 1870.

THE COLONIAL SYSTEM

In the 1800s Equatorial Guinea was still subsidiary to Spanish interests in the Caribbean. Indeed, the budget of the colony was not separated from Cuba's until 1884. Spanish colonialism had little sub-Saharan interest: "Above all, throughout the history of Spanish Guinea, the ignorance of the authorities and of the Spanish people of the value and uniqueness of the Territories of the Gulf of Guinea appears; after the infatuation with the colony of Cuba, from 1898 on, Morocco was preferred to Guinea."[12]

Between the final delimitation of Rio Muni's boundaries in 1900 and the advent of the dictatorship of Francisco Franco in 1936, a full colonial system was installed. Unfortunately for the metropole, it was marked by fits and starts, as well as a lack of administrative continuity. For instance, between 1858 and 1968, there were more than 99 changes in the governorship.

Early in the twentieth century numerous plans were presented for turning the colony over to concessionaires or for selling it to a foreign power. For example, in 1905 the Sociedad Fundadora de la Compañia Española de Colonización asked for exploitation rights. The Sociedad sought to convince potential investors that "our Guinea colonies, by their maritime position, have a great advantage over other African possessions, such as the Belgian Congo or the Sudan where the prices of merchandise suffer great surcharges because of the long distances they have to cover."[13] The group generated little response and turned its interest toward Morocco as a more fertile field for colonial endeavor. Development of the colony was not given to a concessionaire, although the proposal was made in the annual budgets until 1914. Africanist congresses, which met in Madrid in 1907 and 1910, in Saragossa in 1908, and in Valencia in 1910, attempted to keep the colonial flame alive. The hope that something might be done in sub-Saharan African bore fruit in the creation of the Liga Africanista in 1912–1913.

In 1902 the Consultative Council for the Spanish Possessions in West Africa was formed under the authority of the minister of state. It was charged with overseeing colonial legislation, public works, concessions, taxation, and colonization. A governor-general resided in Malabo (Santa Isabel). Early in the century, the administration of the mainland was divided between Bata, where a subgovernor for northern Rio Muni resided and Elobey, where a subgovernor for the southern part of the territory lived. Annobón was entrusted to two minor officials directly responsible to the governor-general.

In 1904 a decree established the basic rules of land concession and designated the colony a *colonia de explotación*, not a European settlement.

Headquarters of the Cámara Agrícola, Bioko. (Photo courtesy Charles W. Grover)

The office of governor and chief of the naval station were separated in an effort to devote greater attention to colonial development. The law also created the Patronato de Indígenas (Native Trusteeship). In 1906, the Cámara Agrícola (Agricultural Chamber of Commerce) was established on Bioko to oversee the interests of the planters. The colonial government itself created the Curadoria Colonial (Colonial Labor Office) to oversee the payment and treatment of labor. Governor-general Angel Barerra, who dominated from 1910 to 1925, strenuously pushed for economic expansion and administrative coherence. The dictatorship of General Miguel Primo de Rivera (1923–1930) promised to give new impetus to the somewhat phlegmatic colonial enterprise. In 1925 the colony was transferred from the purview of the Ministry of State to the Bureau of Morocco and Colonies (Dirección General de Marruecos y Colonias), a department of the Presidency of the Council of Ministers. In 1934 the dirección was abolished by the government of the Spanish republic.

After 1936, the regime of Generalissimo Francisco Franco brought about several administrative changes. Colonial affairs were emphasized and greater stress was placed on continuity of policies and personnel. In 1942 Spanish Guinea was returned to the authority of a resurrected Dirección General de Marruecos y Colonias. In addition, in 1942, Rio

Muni was divided into seven administrative territories dependent on the regional administrations in Bata (the coast), Evinayong (the center), and Ebebeyin (the east). A push toward development and organizational efficiency was especially evident in the administration of Governor-General Faustino Ruiz, who took office in 1949.

In the 1950s Spain's sub-Saharan territories were declared overseas provinces. Although in 1959 Africans were granted metropolitan citizenship, this arrangement was short-lived; the African territories were granted autonomy in 1963. The responsibility for Spain's remaining overseas territories was vested in the Dirección General de Plazas y Provincias Africanas.

Although colonial penetration was slow, the almost nonexistent Spanish presence of the nineteenth century was replaced by a growing number of Europeans. Population statistics for the pre-Franco period are sometimes scanty and inconsistent. In 1907 Spanish Guinea as a whole held 404 Europeans, and in 1910 the colony's capital held 206 whites and 1,815 blacks out of a population of 2,021 inhabitants. In 1916 the whole of Spanish Guinea held approximately 600 Europeans. Sixteen years later, the colony held 1,533 Europeans: Rio Muni had 420, out of a population of 131,125, and Bioko had 1,113, out of a population of 34,204. In 1942 there were 799 Europeans and 134,424 Africans in Rio Muni. There were 1,529 Europeans on Bioko, out of a total population of 23,000. By 1960 the number had risen to 4,222 out of a total population of roughly 63,000 on Bioko. In Rio Muni there were 2,864 Europeans out of a population of approximately 200,000.[14] In 1968 the European presence was one of the strongest in equatorial Africa with relation to the indigenous population. There was one white for each 4 square kilometers, versus one white for 0.2 square kilometers in Cameroon and 0.3 square kilometers in Gabon in 1969.

The increased metropolitan presence saw the slow growth of a colonial apparatus, especially in the capital. An order of 1907 recognized the city council (Consejo de Vecinos) in the capital and charged it with overseeing street lighting, health, public buildings, and other municipal concerns. In 1929 this order was superseded by one that defined the council as the populace's legal representative.

By the 1920s Malabo was already a superficially Hispanicized colonial town dominated by a European community larger than the old Anglophone Fernandino elite. In town, residential segregation continued to be based on socioeconomic rather than purely racial criteria. As in Portuguese-dominated Africa, colonial policy differentiated between assimilated and nonassimilated Africans; in the late 1920s the status of *emancipado* was created. *Emancipados* existed outside the "native" justice system and were subject to European law. In 1944 the degrees of emancipation were further elaborated and *emancipados* were subject to metropolitan law, although this did not excuse them from changes introduced by colonial statutes. In 1948 land legislation inheritable property was recognized only among Africans who had become Christians. The significance or insignificance of *emancipado* status must be realized. In the mid-1950s, out of an African population of over 100,000, approximately 100 were *emancipados*.[15]

Unemancipated Africans were under the jurisdiction of the Patronato (created in 1904) and were legal semiminors. In addition, they were restricted to the ownership of 4 hectares of land. The Patronato could use its control over land as a threat to recalcitrant Africans, for the law stated that native "property will be subject to seizure and encumbrance by the Patronato, or any other credit or colonial agency that may obtain these privileges, by decision of the Council of Ministers." In the late 1930s an attempt was made to base the administration of justice on modified "native usage" (e.g., polygyny, though permitted, was discouraged through a system of fines after the third marriage). A Supreme Native Tribunal capped a hierarchy of native courts. It was composed of Africans chosen by the governor-general and sat under the presidency of European officials. In 1952 the Patronato established the Delegaciónes de Asuntos Indigenas (Delegations of Native Affairs) to supervise the property transactions of the unemancipated. The Patronato was abolished seven years later when all Africans gained metropolitan rights.

Until well into the twentieth century, much of the population remained beyond the influence or control of these European agencies. On Bioko, the Bubi staged armed resistance in 1898 and, in the following year, Moka died. In 1904 Sas Ebuera, Moka's successor, died after resisting further. Malabo, the leader who inherited what remained of the Bubi paramountcy, although amenable to Spain, was not respected throughout the island. It was not until 1910, after the defeat of Luba (a Bubi leader), that the Bubi were finally brought under Spanish control.

Spanish control was even more tenuous on the mainland. In the early 1920s colonial authority was still unestablished, whereas France had long since "pacified" the Fang on its side of the border. Resistance was sporadic because alien intrusion was half-hearted and uncoordinated. In 1906 the Fang of Mebonde, directed by Chief Obama Mbain, captured the Spanish governor himself; the official was eventually rescued through the efforts of a private European hunter. Angel Barerra was the first governor-general to visit Rio Muni extensively. His successor, Miguel Nuñez de Prado, undertook final "pacification" in 1926–1927. Spanish rule tended to be indirect, with local affairs left largely in the hands of traditional rulers or groups. Native courts overseen by Europeans dealt with issues that could not be handled at the local level or seemed to run counter to European morality. A Colonial Guard, composed of Africans, maintained Spanish rule, which was regulated by colonial functionaries scattered throughout the lightly settled territory. Chiefs were often given authority over territorially delimited areas, rather than over ethnically defined groups. "This variation of 'indirect rule' through appointed chiefs is dictated by Spanish inability to gather sufficient military and administrative personnel to institute full, direct control over the indigenous population." As a result, it was "convenient to keep the native peoples in a state of moderate contentment under separatistic, pseudo-traditional forms while the colonial power attempts to weaken sources of potential opposition among certain kin groups and manipulates the strings attached to puppet chieftains."[16]

Bata was the only notable urban center on the mainland. It was, in contrast to Malabo, a much newer town, characterized by far fewer amenities. Before 1900, France had a customs and military post on its site. In 1905 the Spanish established a new settlement 3 kilometers to the north and dismantled what was left of the French buildings. In its early days the city was quite unimpressive. In 1907 the African population numbered 200, the naval barrack was a shack, and the quay was in a state of deterioration due to the surf. The twenty-five-member Colonial Guard was under the authority of a single European officer.

The fragile nature of colonial rule in Rio Muni, in comparison with neighboring Gabon, even caused some migration into the Spanish territory to escape the *corvée*. In the 1920s and 1930s the construction of the Congo-Océan Railway in the French Congo drained a large area of manpower. The Fang referred to such recruitment as "the terror" and used any means to escape. Spanish Guinea was viewed as a magnet for the border population because it was a place "where censuses, native taxation, levies and native justice are unknown." According to one French official, it was possible for Africans to live "in complete freedom" in Rio Muni.[17] A further inducement to flight was the relatively enlightened labor policy of the Spanish republic of 1931. On the other side of the border, a French colonial officer noted, "At the time of my arrival not a soul was in the village; everyone under the reign of the Terror has fled into the bush or into Spanish Guinea."[18] In the 1940s the administration in Gabon noted that "at the slightest provocation, the natives leave their village to become vagrants, or even seek refuge in Spanish Guinea." In 1956 a U.S. visitor observed that "save for missionaries and government personnel who have penetrated the interior or established posts along the colony's periphery, European activity has been limited to the coast, particularly to the towns of Bata and Benito [Mbini]."[19]

THE COLONIAL ECONOMY

If the administrative establishment of Spanish colonialism was slow, the development of a *colonia de explotación* was even more halting. Spanish colonialism in Rio Muni and on Bioko had one thing in common: a lack of physical and financial infrastructure.

The Physical Infrastructure

On Bioko, it was often very hard to get cocoa to port for shipment. In 1911 a projected 190-kilometer (118-mile) railroad was begun. The first segment ran from Malabo to Basupu on the eastern side of the island, but no further segments were completed. By the 1930s even the initial segment had been abandoned, a product of fluctuating world market prices and a dearth of labor.

Roads. Motor roads played a more prominent part in Spanish strategies in Rio Muni. In 1926, a network was started; 200 kilometers of roads were constructed. There were 750 primary roads; of these, 250 were asphalted. The principal axis of the Rio Muni road system follows the Cameroon border and was used for Spanish penetration in 1926 and 1927.

Forestry. From the colonialist point of view, Equatorial Guinea was undercapitalized and underexploited. In 1928, one hundred and fifty years after Spain claimed possession, only 9.2 to 12 percent of Bioko's cultivatible land was used. In Rio Muni, only 0.026 percent of the cultivatible area was in use for cash crop cultivation.[20] This difference has had significant ramifications down to the present time. Forestry is not labor-intensive and, perhaps as a consequence, became the chief form of colonial economic activity in Rio Muni. The Compañia Forestal de Benito tried lumbering in 1919 and employed mainly non-Spanish Europeans in managerial tasks. A decade later the Cámara Agrícola y Forestal de la Guinea Continental Española was formed to coordinate forestry and agricultural interests. Initially, development costs seemed prohibitive to many would-be foresters. Eventually, concessionaires were granted most of the timberland eastward from Bata to Niefang and southward from these two towns to the Gabonese border. Several of the forestry lands granted were more than 20,000 hectares.

Plantations. Outside the Bata–Sevilla de Niefang–Gabon border area, tropical exploitation was more difficult. To the north, with the exception of the Ntem, there were no navigable rivers on which to transport logs. The area to the east of Niefang was inaccessible. In 1920 there were 14 trading houses in Bata and 2 or 3 in Mbini. The islet of Elobey

Chico also attracted a number of traders, drawn by the supposedly salubrious climate. Besides lumber, Rio Muni produced, in order of importance: cocoa, coffee, and palm oil. By 1930 cocoa plantations had been established along the Bata-Ebeyiyin road. Coffee was also developed by returned Fang migrants from Bioko and, at the end of the 1930s, the territory produced 245 metric tons. Large palm oil plantations were set up with the aid of mechanical presses and the adoption of Asian oil palm plants. One company that had particular success was Sociedad Colonizadora de Guinea (SOCOQUI), which exploited palm oil at Cape San Juan.

Bioko was the cynosure of Spanish colonial endeavor. After World War I, the island experienced a spurt of growth. Some new economic activities were introduced and remained totally dominated by Europeans. For instance, cattle ranching flourished in the uplands. In the mid-1920s the Compañia Nacional de Colonización Africana (ALENA) acquired most of the ranch property held there by the Compañia Transátlantica. Later large areas of the highlands were taken for other European ranches and farms.

In agriculture, the Fernandinos faced increasing competition from European plantation owners and trading companies. In 1912 the colony already contained 35 Spanish firms. The economic position of the black planters made them continued prey for European speculators and competitors. Beyond their insolvency, their nationality was a factor held against them. The Spanish administration increasingly questioned the wisdom of allowing black settlers to establish farms. By the beginning of the 1920s, a winnowing process was well underway. Lack of facility in Spanish made winning land disputes extremely difficult. Those members of the black planter group who could not adjust often migrated to British West Africa.

In addition to human impediments, the black planters were confronted with a fluctuating and unpredictable monoculture. The island had suffered neglect in the nineteenth century, which bequeathed a meager infrastructure needed for continued and expanding exploitation. Bioko contrasted strongly with São Tomé where many of the plantations had their own systems of small railways. The colonial regime had hardly taken an interest in linking the capital with other centers of settlement by the opening decades of the twentieth century. In addition, cocoa farmers could seldom cooperate to any effect. Legislation increasingly divided them along ethnic and economic lines. The budget of 1911 established a quota system for cocoa that granted a rebate on the duty paid on the first 2,000 metric tons of cocoa imported to Spain. This arrangement benefited the larger proprietors, but many small black

Abandoned main house of a European *finca*, Bioko. (Photo courtesy Charles W. Grover)

farmers could not manage to get their cocoa included in the quota and had to dispose of it at less advantageous terms.

A new labor ordinance was promulgated in 1913. Its conditions were especially onerous for the small planters. One provision stated that laborers were to be taken away from insolvent planters. When all factors combined with the colonial bias toward European-owned *fincas* (plantations), it was only a matter of time before the Fernandinos gave way to the newer and larger European units. Small planters increasingly fell into debt to Spaniards, although Maximiliano Jones (1870–1944), the richest individual on the island, remained a symbol of the former prominence of his community.

In 1928, approximately 98.6 percent of Bioko's exploited land was devoted to cocoa.[21] Production was hampered by a lack of labor, lack of credit facilities, and poor infrastructure. It was obvious that the Cámara Agrícola was not effective in encouraging colonial agriculture. Most of the large Spanish proprietors lived in Barcelona, the cámara's metropolitan headquarters. Few were willing to delegate authority to their agents. In 1923 the Unión de Agricultores de la Guinea Española was organized, with aims somewhat more precise than those of the Cámara Agrícola. It was to seek the introduction of new products, stimulate consumption of colonial produce, harmonize the interests of different segments of the cocoa industry, and provide agricultural credit to members. Five years later, the Sindicato Agrícola de los Territorios Españoles del Guinea was formed. It included owners, renters, administrators, and usufructuaries, and made a special provision so that female heads of families could be members. The sindicato was empowered to issue stock, a policy different from that of the Cámara Agrícola.

Although many persons on Bioko entered cocoa growing, production was less than that of smaller São Tomé. As mentioned, topography impeded some land clearing. In addition, the cultigen itself presented certain difficulties. One type of cocoa, Amelonado, does best in virgin soil and takes several years to mature. Farmers had to clear land and then wait seven years for the first trees to produce. Thus, they had the difficult task of maintaining themselves for several years before receiving any return on their investment. In addition, neglected plots deteriorated rapidly. It was unprofitable to begin cultivation unless it was fairly definite that there would be available laborers. Also, perhaps in an attempt to maximize production, many farmers cleared away shade trees from their cocoa farms. This exposed the cocoa plants to insects, especially thrips. Cocoa was especially subject to black-pod disease. Because of the hilly terrain, the plants had to be hand-sprayed with a mixture of copper sulphate and lime.

The Financial Infrastructure

The organization of colonial finances was a perennial problem. Governor-generals frequently talked of the need to deposit colonial revenues in a Spanish bank. Nevertheless, the Bank of British West Africa was the chief bank until the Franco era. Attempts to establish a Spanish bank foundered. In 1916 a Banco Colonial Español del Golfo de Guinea was formed with the blessing of the government, but encountered opposition in the Spanish Parliament. The project collapsed and was not revived until after 1923. Various banks were contacted in an effort to provide a bank for the Spanish colony. In 1930 an agreement with the Banco Exterior de España gave it charge of the colonial treasury. Unfortunately, this move came at a time of worldwide economic distress. The bank did not begin its functions and, in 1932, the project was abandoned.

The colonial regime encouraged the entrance of metropolitan capital and the creation of a protected market in the metropole. A 1905 land law encouraged the influx of European proprietors. The law gave the governor-general the right to confer deeds to plots of undeveloped land not permanently claimed by the state. The governor-general's right to confer such grants was confined to areas not exceeding 100 hectares or

1 square kilometer. The president of the Council of Ministers in the metropole could grant from 100 to 1,000 hectares. The approval of the government as a whole was necessary for a grant of 1,000 hectares. Land transferred from state control to private or corporate ownership was tax free for a period.

By 1925 cocoa production had increased 143 percent over 1910.[22] Spanish colonialists were able to argue that at last the colony was useful. Bioko produced more in customs revenue than it received as a subvention from Madrid. According to its boosters, the amount received was undervalued because it was paid in gold. A call for protection for colonial cocoa was based, in part, on this fact. It was argued that colonial produce was cheap and complaints against it were the result of price gouging by chocolate manufacturers. Colonial interests further noted that in 1918 Bioko ranked third, after the Philippines and the Canaries, in terms of Spanish imports. Imports from the Bight of Biafra were far higher than those from Rio de Oro or the Spanish zones in Morocco. In 1918 Bioko was seventh among the Spanish territories in terms of its exports to the metropole, a situation that colonialists laid to the disruptions of World War I. Spain imported more from Bioko than it did from former colonial areas and areas of semicolonial penetration, such as the Philippines and Algeria. In terms of import and export duties collected, Bioko stood well ahead of the Canaries and the Philippines.[23]

Colonial interests agitated throughout the 1920s for a reduction in duties. In 1928, after several years of deliberation, Madrid liberalized its import policy. Some products were exempted altogether. Duties on coffee, coconut palms, and wood planks were reduced. The duty on cocoa entering Spain was not reduced, but that leveled on cocoa exported elsewhere was cut. The export of rubber to the metropole was taxed only to a minimal degree and imports on cola and oily almonds were considerably reduced.

In spite of great progress, the record of the 1920s is mixed. In 1929, exports were almost exclusively Bioko cocoa, plus some wood from Rio Muni. Six million kilograms of the cocoa were sent to Spain each year. Ten thousand tons of wood were sent from the colony, mostly to Hamburg because of the paucity of Spanish shipping. In the year of the Great Depression, the economic outlook did not presage the tremendous changes that would take place in the following thirty years.

Interestingly, the opening years of the Depression did not drastically affect the export economy of Bioko. During the Spanish Civil War (1936–1939), economic disruption was caused initially by struggles among the European population. Bioko declared for the nationalists of Generalissimo Francisco Franco, while Rio Muni remained with the republic. Moroccan troops were sent in support of the Francoists and Rio Muni was subdued by October 1936. The importance of the colony to the war in the metropole is indicated by the fact that it contributed almost all of the nationalists' crude lumber and almost half of their copra and palm oil.

In the wake of the civil war, the colony's economic situation began to alter. Spain's neutrality during World War II may have, on the whole, benefited the economy. Spokesmen for the regime liked to maintain that colonial "prosperity" was the outgrowth of the authoritarian precepts of the government. Statist economic policies, along with oligopolistic manipulation of cocoa and coffee prices, assured invested Spanish capital a handsome return and transformed the previously marginal indigenous population into an independent and largely complacent group of cash crop farmers.

The impact of the war was also felt in 1940 when Franco sent 2,500 Spanish and Moroccan troops to Bioko, supposedly to thwart any British attempt to take the island. The war cut down on shipping between Spain and its colony, which forced some readjustments. Cocoa planters were able to shift to coffee and oilseed production, and rubber production stepped up. The okume wood of Rio Muni, which had been used by the *Luftwaffe*, was allowed to grow. At the same time, forestry companies shifted to the use of less valuable lumber for sale on the Spanish market. By the end of the war, almost three-quarters of Spain's coffee came from Spanish Guinea. Small quantities of dried manioc and bananas were also exported.

In overview, the period is one in which colonial imports increased dramatically. From 1932 to 1934 only 3 percent of the metropole's imports came from the Canaries and the African colonies. From 1940 to 1946 their percentage rose to 16 percent. In percentage terms, the height of colonial imports was in 1942. Spain's small empire accounted for 21 percent of its imports. The Canaries, which were listed as a colony for the purpose of trade statistics, contributed half of these imports and Spanish Guinea most of the rest.[24]

In Spanish Guinea the pace also quickened. By the early 1940s significant change had taken place in Bata itself. Although the municipal infrastructure still lagged far behind Malabo, the settlement was truly a town. The Banco Exterior de España had a branch there, which served as the center of Rio Muni's European business community. There was also a colonial treasury, post office, agricultural service, customs, and public works offices. Trading posts boomed. In Bata before 1936 there were 7 Spanish and 12 foreign *factorías*, of which 6 were Lebanese. From 1937 to 1941, 11 Spanish and 12 foreign traders established themselves.[25]

Franco's Spain recognized the value of colonial Africa because of the prestige derived from both a continued place among the colonial powers and more tangible economic benefits. In 1954 the director of the Dirección General de Marruecos y Colonias said that "without doubt the Guinea territories have a special economic interest for the metropole, for they provide us with certain exotic products that otherwise would demand considerable investments."[26] The colonial economy was molded by three groups: forestry concessionaires, resident Spanish *fingueros* (plantation owners), and metropolitan-controlled plantation companies. From the mid-1930s on, economic development was overseen by marketing syndicates that conformed to the corporatist ideas of the regime and guaranteed a higher than world price for planters and bankers. *Sindicatos* empowered by the government controlled the marketing of cocoa, coffee, and lumber. A cocoa syndicate, a coffee syndicate (PRO-GUINEA), and a forestry syndicate were each run by a central committee based in the metropole. The committees were almost entirely composed of board members of companies involved in the exploitation of cocoa, coffee, or timber.

As early as 1930 it was apparent that Spain intended Bioko (especially its uplands) for use by European cultivators. At that time 18,000 hectares had been conceded to Africans, while some 21,000 hectares had gone to Europeans, a situation that remained legally frozen until 1948. In 1942 and 1943, out of 40,000 hectares devoted to coffee and cocoa, only 4,000 were in African hands[27] (see Table 2.1).

The decreasing availability of exploitable land meant that oligopolistic control had already solidified; membership in the ranks of landholders was likely to drop rather than rise, because larger units tended to buy out smaller ones. All cocoa producers on Bioko were obliged to join the Cámara Agrícola. Approximately 65 percent of the annual cocoa harvest was consumed in Spain, where the syndicate controlled its importation and distribution. The price of cocoa was fixed by the syndicate, which warehoused the crop in both colonial and Spanish ports. The organization also had a monopoly on imports to Spain.

The formation of syndicates muted the potential for conflict between the agricultural interests of Bioko and the lumber interests in Rio Muni. In 1936 a timber syndicate (Sindicato Maderero) was created in Bata. In the following year the corresponding metropolitan body, the peninsular delegation of the timber syndicate, was formed. Unlike the cocoa syndicate, the forestry organization did not depend on inflated metropolitan prices for protection. Instead, it relied on highly protectionist tariffs, which effectively eliminated foreign competition.

By the late colonial period, most of Equatorial Guinea's products were protected in the metropolitan market. In the 1960s the Syndical Committee for Cocoa fixed the price of the crop at approximately 150

TABLE 2.1
Export of Cocoa, 1894–1972 (tons)

	Province		
	Bioko	Rio Muni	Total
1894	—	—	500
1900	—	—	900
1910	—	—	2,445
1920	—	—	4,741
1930	—	—	11,606
1939–1940	—	—	14,665
1949 1950	15,759	1,857	17,616
1953–1954	16,548	1,412	17,960
1954–1955	20,039	1,460	21,499
1955–1956	18,116	2,070	20,186
1956–1957	21,529	1,944	23,473
1957–1958	19,554	2,271	21,825
1958–1959	20,971	711	21,682
1959–1960	25,433	2,477	27,911
1960–1961	22,100	3,250	25,350
1961–1962	23,559	2,400	25,959
1962–1963	28,673	2,835	31,508
1963–1964	29,458	3,898	33,356
1964–1965	31,305	3,435	34,740
1965–1966	31,014	4,493	35,507
1966–1967	35,344	3,331	38,675
1967–1968	30,058	3,506	33,564
1968–1969	36,340	2,651	38,991
1969–1970	—	—	26,943
1970–1971	—	—	28,384
1971–1972	21,264	3,154	24,418

Source: Armin Kobel, "La République de Guinée Equatorialle, ses ressources potentielles et virtuelles. Possibilités de développement," Ph.D. diss., Université de Neuchatel, 1976, p. 267, citing Resumen estadística del Africa española, 1932–1960. Reprinted by permission of the Bibliothèque Centrale de l'Université, Université de Neuchatel.

percent of the going world price. In 1966 the colony was the fifth largest continental grower after Ghana, Nigeria, the Ivory Coast, and Cameroon. Production was about 3 percent of the world total.

Colonial lumber was shielded by a duty on foreign lumber, which reached 17 percent in 1967. The tariff assured that wood from Spanish Guinea would have a favored place in metropolitan markets. The premium paid on coffee from Spanish Guinea put it well above the international market price. This premium gained an extra 240 million pesetas or approximately $3.5 million for Spanish producers. Spanish coffee companies were forced by law to buy quotas of colonial coffee at inflated prices before they were permitted to buy cheaper foreign produce.

By the 1960s the economy of Spanish Guinea was dominated by 11 major groups: 4 companies, 2 banks, and 5 plantation-owning families. The 4 companies produced 30.3 percent of the annual coffee crop and 9.5 percent of annual cocoa production. The 2 banks produced 4.6 percent of the annual coffee crop and 20.1 percent of annual cocoa production. The 5 major families produced 30.3 percent of the annual coffee crop and 29.9 percent of annual cocoa production.[28]

MIGRANT LABOR

Colonial prosperity rested upon an extensive system of government-regulated migration. A continuing theme was the desperate need for manpower and uncertain sources of supply.

Bioko had a sparse population and was, at the same time, the focus of the colonial economy. The "decadence" of the Bubi may have been exaggerated by certain colonial officials, but their proportion of the total population definitely was declining. Various nineteenth-century epidemics, including yellow fever in 1868, smallpox in 1889, whooping cough in 1893, and dysentery in 1896, were probable causes. Trypanosomiasis was spread on the eastern side of the island by migrant laborers from the mainland.

Some officials looked to Rio Muni as a source of manpower because of the demographic decline on Bioko. The Spanish conquest of the interior in the 1920s encouraged the idea. However, such migration proved insufficient because of the lack of firm Spanish control over the territory, combined with the low population density and the resistance of its inhabitants. The introduction of coffee production in Rio Muni in 1926 was also a disincentive to the outflow of labor. The Cameroonian coast, to the north of Rio Muni, would have seemed the most likely recruitment area. The demographics of coastal Cameroon did not favor the large-scale outflow of labor, however. In contrast to Eastern Nigeria, which had a density of 1,000 people per square mile in the central Igbo region in the 1920s, the Kumba area of Cameroon reported a density of only 15 to 20 people per square mile.[29]

Labor from British West Africa and Liberia was more expensive but more plentiful. After 1900 even workers from these areas were harder to procure. A Spaniard sadly noted the situation: "Twenty-years ago the number of colonies in the Gulf of Guinea was very few and much less the number of plantations; today, however, all this has changed and there scarcely exists a mile of beach where the flag of some European nation does not fly, and where the black labourer is not a necessity."[30] In 1900, a large number of agricultural workers from Western Nigeria went on strike and some Europeans feared a general colonial insurrection.

The Spanish sent military reinforcements and the incident brought forth denuncations in the British Parliament. Four hundred fifty workers from Nigeria were repatriated at Spanish expense. The strike expressed the willingness of the workers to mobilize on an interethnic basis for the protection of their common interests. It also gave impetus to calls for a labor embargo from British West Africa, an action that only served to increase dependence on increasingly coercive labor recruitment elsewhere.

British interest in labor conditions in Spanish Guinea was a by-product of the furor produced over conditions on São Tomé and Príncipe. The former was the world's largest cocoa producer. One outcome of this tremendous push to produce was labor abuse. As early as 1902 rumors of deplorable conditions caused grave concern. British protests finally resulted in a cocoa boycott. Among other things, the British feared that Sierra Leoneans were being exploited in a carry-over from the slave trade. Moreover, such migration was a drain on already insufficient colonial manpower.

Between 1910 and 1914 the British investigated labor conditions on Bioko. The Spaniards found themselves trying to increase production while fending off labor investigation. As a result of British probes, foreign enlistment ordinances were issued by Sierra Leone, the Gold Coast, and Southern Nigeria. Various colonial governments promised to cooperate against illicit labor traffic. A resident vice-consul was placed on Bioko. In spite of these measures, there was continuing evidence of illegal migration.

Around 1904 the Spanish colonial regime expressed faith in the future, but recognized present realities. Internal supplies of colonial labor should be tapped, but the limitations of this source were recognized.[31]

The government noted that there were 993 laborers from Liberia on the island, versus only 256 from mainland Spanish Guinea.

In Liberia the government had long connived at the forcible recruitment of labor from the Kru, Grebo, and Vai peoples. In 1905 Spain agreed that it would stop paying each laborer and instead pay the Liberian treasury in gold. In an effort to procure more labor, attempts were made to improve conditions. In 1906 a Native Labor Code (Reglamento del Trabajo Indigena) was put into effect. The measure did not dramatically increase the flow of labor. Indeed, in 1908 a Liberian law restricted the areas of labor recruitment. The following year, two Spanish officials visited Liberia in an effort to procure a regular flow of labor.

In 1914 Spanish Guinea and Liberia completed a labor convention, subject to termination by either party on six months' notice. Between 1919 and 1926 a known 4,268 laborers were legally recruited and employed. It was estimated that the total number of laborers from 1914 to 1927, when Liberia terminated the agreement, was at least 7,268. Throughout its lifetime the agreement seemed precarious. Spanish ships were unwilling to land workers at any other place except Monrovia. There was also a feeling that the agreement deprived Liberia of labor.

The shipment of labor was difficult in the years immediately following 1914. As a result, a 1918 decree authorized the reengagement of workers who had completed their contracts for an additional two years. World War I threatened to halt the movement of labor to neutral Spanish Guinea from neutral Liberia. The Spaniards were accused of letting ammunition into German-held Cameroon and, in late 1914, an Anglo-French blockade of Bioko and Rio Muni began. Spanish Guinea's plantation economy seemed in danger of collapse. Instead, with the fall of the German colonial regime in Cameroon, 16,000 soldiers and followers fled to Bioko. Although the British feared the construction of a German redoubt in Spanish Guinea, their fears proved unfounded. The influx gave the island a new and unexpected infusion of labor. About 5,000 to 6,000 of the refugees from Cameroon were African troops. Until the end of the war their presence temporarily helped alleviate the acute labor shortage.

In the aftermath of the war, the agreement with Liberia went into high gear. However, as before, the British remained concerned about illegal migration of Africans from Sierra Leone via Liberia. Early in 1923 they unsuccessfully tried to get an agreement from Spain against illegal labor recruitment. Such efforts continued throughout the middle 1920s. In 1925 Liberia itself temporarily stopped the shipment of labor after a diplomatic incident. As an upshot of the affair, Liberia demanded £500, which was collected by the island's planters. No Liberian laborers were sent in 1925 and only forty arrived in the first six months of 1926. Government statements from Monrovia gave little indication of a willingness to cooperate. In March 1926 the president of Liberia visited and seemed to hold out the promise of business as usual. The next year, however, matters came to a head when Liberia announced the cancellation of the labor agreement.

Once again the island's economy appeared to be on the verge of collapse. Only 80,000 hectares were under cultivation and it was estimated that 40,000 Fang would be needed to put the maximum amount of land into cocoa production. The uncertain nature of the labor supply caused the government to go far afield in search of alternatives. In 1927 the idea of colonizing the island with Asian workers was proposed and early in 1928 the governor-general suggested that workers for the lowlands be sought in China or Malaya. In desperation cocoa planters turned an eye to Portuguese Angola and Mozambique. Around 1930 the Spanish consul in Monrovia reported that the island would receive an adequate supply of Japanese labor.

Liberian labor remained available, however, through increasing coercive methods of recruitment. In early 1928 a private agreement was concluded between the Sindicato Agrícola de Guinea and a group of Liberian politicians. Between autumn 1928 and the end of 1929, 2,431 workers were sent.[32]

In 1930, the League of Nations launched an investigation of the traffic. Although the league did not censure the Spanish administration, labor conditions appeared harsh. One half of the worker's salary was paid in Spanish currency on Bioko. The other half was supposedly paid on return to Liberia. The laborers were usually illiterate and rarely kept proof of payment. Many returnees complained of insufficient payment or none at all. Also, when laborers changed employers on the island, they were paid only by the last employer for the last employment period. In 1931, a African-American journalist visited and reported that conditions were still deplorable:

> In addition to the weekly rice-and-fish ration, they received also a kilo of coffee and a cup of palm oil. They revealed that on Fernando Po [Bioko] they were put to work at 6 a.m., worked until 11 a.m., went to work again at 1 p.m. and quit at 6. . . . They lived in warehouses, fifty "boys" being packed close together on beds of cocoa staves and banana leaves. Women were difficult to get and those available were diseased. If the "boys" contracted sleeping sickness, venereal disease or any of the other numerous maladies to be caught there, the Spanish sent them to the hospital, but they received no pay during their illness, whether or not they were at fault. These conditions prevail in Fernando Po [Bioko] today.[33]

The rapid increase in colonial development that characterized the late 1920s continued into the 1930s. There was repeated evidence that labor was smuggled in from British West Africa and Liberia. As in the past, labor shortage did not result in the amelioration of conditions but instead increased the various subterfuges to lure labor and detain it beyond the terms of its contract. Women and minors were recruited and vagrancy laws were employed to obtain plantation labor. Spain adhered to the Forced Labor Convention of 1930, but forced labor (*prestación personal*) was not abolished until the late 1930s.

In 1933 the French government complained to the League of Nations about the treatment of workers lured from Cameroon. A year later the Spanish governor-general and the French commissioner for Cameroon signed a labor agreement permitting the annual recruitment of 4,000 laborers. The agreement did not prove satisfactory to the French and, in early 1936, it was abrogated.

The "boom and bust" experience of the 1920s made the Spanish government extremely anxious to insure an adequate flow of labor through intergovernmental agreement. In 1939 the British began a study of labor migration to Spanish Guinea. Three years later an agreement was signed governing the recruitment and treatment of Nigerian laborers. It provided a maximum six years of contract and gave the British consul the right to inspect labor on plantations. A labor officer in Calabar was assigned to oversee the agreement. British authorities permitted the recruitment of up to 250 laborers per month.

In spite of licit and illicit labor trafficking to Bioko, the island's planters continued to complain of a labor shortage, just as they had before 1930. In 1944 a delegation from the Cámara visited Calabar to arrange an increased rate of recruitment. The Nigerian colonial authorities, for their part, continued to receive reports of labor abuse. The Guardia Colonial, composed largely of resident Nigerians, was accused of rampant brutality. Employers could have laborers flogged and placed in prison for indefinite lengths of time. The Spanish administration showed little interest in abiding by the repatriation terms of the labor agreement and was accused of being lax in responding to complaints from the British labor officer. Social conflict was engendered by the shortage of women.

In 1941 there were 10,000 Nigerians already on Bioko. This number increased throughout the decade. In 1954–1955 a very conservative estimate of the total number of Nigerian migrants on the island was "about 15,800."[34] In the mid-1960s there were approximately 100,000 people on Bioko, of which about 85,000 were Nigerians; two-thirds of these Nigerians were Igbo-, Ibibio-, and Efik-speaking.[35] Several reasons for this outflow can be adduced. First, recruiters were generously paid by Spanish officials and employers. Second, there was pressure to pay taxes to the British colonial administration in Nigeria. And, third, there was demographic pressure; labor came from the most thickly populated areas of the Eastern Region. Thus, it was part of the more general Igbo diaspora that sent the Igbo and related peoples into northern Nigeria, Cameroon, and beyond.

In 1950 the labor agreement was revised to conform to the conventions of the International Labor Organization and provision was made for the repatriation of illegally procured labor. Three years later the Nigerian minister of labor headed a delegation to Bioko to investigate persisting charges of labor abuse. Although he failed to find evidence of substantial abuse, the minister was able to arrange an increase in the rate of pay and the registration of all workers, both treaty and nontreaty. A new agreement signed in 1954 included these improvements and also established a fund to improve the educational and welfare services available to the migrants.

In early 1956 a new Nigerian minister of labor headed a delegation, which included the Eastern Nigerian ministers of labor and welfare. Their tours produced a further pay increase of 25 percent, plus an increment in the number of Nigerian workers who could be legally recruited (600 to 800 monthly). It was agreed that compensation was to be paid by the employer to workers or their families for any work-related injury. Other terms of agreement were the elimination of penal sanctions for breach of contract and the payment of a head tax to the Nigerians. The Spanish authorities agreed to pay £3 to the Federal Nigerian government and £2 to the Eastern Region.

In 1957, another group of Nigerian investigators visited and reported some progress. Among other things, the Spanish had opened a primary school and an orphanage for Nigerian children. Nevertheless, the group's report on the cocoa farms was mixed. At best it found prosperous plantations where living quarters were commodious and rations adequate. At worst, it cited farms where eight unmarried men frequently shared a room eight by ten feet and where two or three married couples often shared quarters. The Nigerian report absolved the Spanish authorities from blame in cases of labor abuse. It also said that the Spanish government had sought to punish its nationals who had violated terms of the labor accord. However, the Nigerians recommended that more labor officers be stationed on the island.

In spite of serious and continuing complaints, in early 1959, Nigeria signed an additional agreement with Spain permitting the recruitment of 2,000 extra laborers for a three-month period. In 1961, after the shooting of four Nigerians in Rio Muni, the Federal Nigerian government immediately lodged a protest with Spain. Another delegation was sent and further refinements were made on the 1956 labor agreement. The reworked agreement included provisions for payment of compensation in cases of work accidents resulting in permanent or partial injury to nonagreement workers and prohibited long-term criminal detention without trial. In addition, laborers were no longer obliged to carry passes.

THE SOCIOECONOMICS OF DEVELOPMENT

Labor migration allowed Spanish Guinea to "develop." Some Africans, as well as Europeans, profited from the highly managed colonial economy. Unfortunately, this colonial development only heightened economic differences between the two parts of the colony. These economic differences developed political overtones at the time of independence.

On Bioko, employers used a migrant wage labor force greater than the European and indigenous population combined. Paternal nurturance constricted the Bubi and, at the same time, promised economic benefits. A policy of concentrating the Bubi in fewer villages coincided with colonial laws that deprived Africans of much of their traditional land. The 1948 property code solidified the dominance of European capital. The law did not recognize automatic native ownership beyond the individual's immediate home and garden, except for collective ownership derived from membership in cooperatives and syndicates. The resettlement and sparseness of the Bubi population made it comparatively easy for Spanish agricultural interests to gain possession of the land.

By the 1960s up to 610 meters in altitude of the narrow east, west, and north coasts were almost entirely devoted to European plantations; in 1964, 600 non-African plantations occupied about 36,423 hectares (on the average about 61 hectares per plantation) and African farms occupied 40,000 acres (on the average 5.7 hectares per farm).[36] Other land-use statistics indicate that 55 percent of the island's cultivators controlled less than 3 percent of the crop-producing land in 1962, while 2.3 percent of the farms controlled 53 percent of the cultivated land.

At the close of the colonial period there were 1,142 cocoa farms of less than 10 hectares, 242 of 10 to 30 hectares, 124 of 30 to 100 hectares, and 100 larger than 100 hectares. The 1,608 plantations covered 29 percent of Bioko's surface.[37] Bubi farmers were often persuaded to exchange their plots for less favorable ones to allow the Spaniards to amalgamate farms into large plantations. In return, the large Spanish combines—the *casas fuertes*—often paid a small annual pension to the Bubi and sometimes provided scholarships for their sons to obtain secondary or higher education. Almost all of the Bubi and Fernandino agriculturalists were in debt to large Spanish commercial firms. The Bubi were gradually accommodated to Spanish culture and became vital supporters of colonial rule.

Colonial land policy, which greatly restricted an African's property rights, combined with intensive missionary Hispanicization to encourage agricultural cooperatives and the quest for *emancipado* status. *Emancipados* could own freehold land and/or become leading members of cooperatives. The first cooperative was founded near Moka in the late 1930s with the aim of acquiring seeds, fertilizer, and tools and organizing the sale and distribution of harvests. After 1945 there were cooperatives in, among other places, Moka (European vegetables and poultry), Batete (cocoa, palm oil), Baho Chico (cocoa, palm oil). After 1952, the Delegaciónes de Asuntos Indigenas oversaw lands, the buying and selling of harvests, and the making of loans against native lands.

With the abolition of the Patronato in 1959, members of cooperatives enjoyed full shareholder rights. The organizations remained under the control of the Ministry of Agriculture. As Nigerian laborers became available, the cooperatives contributed more to the total export crop. In 1964 and 1965, companies based in the metropole were responsible for 36.5 percent of the cocoa crop, while 42.8 percent was produced on lands controlled by Spanish family concerns or resident farmers. The residuum (20.7 percent) was largely composed of the harvest from Bubi cooperatives. It is also significant that, by the 1960s, one of two blacks sitting on PROGUINEA's committee was Enrique Gori Molubela, a chief molder of Bubi political opinion and head of the Bioko branch of the Territorial Union of Cooperatives.

The situation in continental Guinea was quite different. In 1962 Rio Muni had only four cooperatives with a total membership of 2,622.[38] Although at the time of independence, Rio Muni had a population of approximately 200,000 (versus 62,612 on the island), timber was its only major export (250,000 tons of okume wood in 1968).[39] Its population of fishermen and peasants had an annual per capita income of $40, compared to $250 to $280 on Bioko and Annobón. In the 1960s only 3.4 percent of the mainland was cultivated for cash crops, compared with 24.4 percent of the island. Infrastructure and educational services were rudimentary and Spanish military rule often brutal. In 1960 agriculture employed 83 percent of the wage laborers on Bioko versus only 25 percent of the wage laborers in Rio Muni.[40] In forestry, cocoa, and coffee activities, the work force was largely imported. In 1962, 68 percent of the colonial labor force in Rio Muni consisted of migrant laborers (6,871 Nigerians and 518 others, mostly Cameroonians and Gabonese).

If economic penetration of Rio Muni was slow, the creation of rural cash crop farmers was even slower: "If it is assumed that all the non-European public labor force was [Equatorial] Guinea's and that the 100,000 subsistence farmers were Guineans, then by 1962 Spanish colonization had brought jobs to 7.1 percent of the employable Guinean population and had given 81.5 percent of the available jobs to non-Guineans."[41] Although Rio Muni had a greater land area than Bioko, the amount of land devoted to the chief export crops was proportionately much less. The output per hectare also varied between the island and the province. In 1955, 625 kilograms of cocoa per hectare were produced on Bioko. On the mainland, 350 kilograms per hectare were produced in the same year.

On both Bioko and in Rio Muni, public sector jobs employed a number of Africans, although they were not proportionately well represented throughout the occupational hierarchy. In the 1960s, 13 percent of the public sector jobs were held by Europeans and 87 percent were held by Africans. Interestingly for future developments, 2,097 (or 60 percent) of the public employees worked in Rio Muni, while 1,339 (roughly 40 percent) worked on Bioko.[42] The wages earned by Africans in that year revealed a significant gap between the migrant laborers the top levels of African salaried workers.

Although Rio Muni lagged behind Bioko in terms of economic development, Spain attempted to give it enough social services to keep it loyal to the metropole. In the 1960s a government housing plan provided for 4,000 houses in three years. The state set up 11 model villages and encouraged the development of a bus service that covered all corners of the territory. In 1963 the Three-Year Economic Development Plan was introduced. It sought to raise the per capita income to $170 per annum and to achieve its more equitable distribution. The three-year plan was also designed to train African personnel, to make essential products competitive on the world market, to generate light industry, to improve credit facilities, and to foster agricultural diversification.

In spite of these measures, there was increased restiveness among those very groups that the colonial government was grooming for power-sharing. Late colonial propaganda portrayed colonialism as providing careers open to qualified applicants. It also defined itself as nonracist. Education was held out as the key to *emancipado* status, which in theory gave complete equality; in practice, this was far from the truth. Social contacts between whites and blacks were rare. Until the extension of Spanish citizenship to all Africans in 1959–1960, interracial sexual relations were tacitly forbidden and sometimes ruthlessly punished. Unemancipated Africans were legal children and "people of mixed ancestry who [were] not acknowledged by the white parent [were] treated as colored wards." Spaniards were "prohibited from performing manual labor or at any time assuming a position subordinate to a native; the illusion [was] steadily promoted that metropolitans [were] demi-gods."[43] Racial stereotypes and epithets were commonly used by the resident European population; *emancipados* were not immune from them. Salomé Jones, an early and active nationalist with wide contacts in West Africa, lamented "the injustice and discrimination that was being practiced in the General Hospital, where the only guarantee [for the best care and treatment] was white skin, because the money of the Negro has no value."[44]

Paternalism persisted. In 1962 Admiral Luis Carrero Blanco, head of the Dirección General de Plazas y Provincias Africanas, announced that state expenditure was higher for the whole of Equatorial Guinea than it was in the metropole: 1,825 pesetas per capita versus 1,800 in the metropole. Also the overall mortality rate was 6.3 per 1,000 in the colony, versus 8.7 per 1,000 in Spain. According to the Spanish officials, the relationship between Spanish Guinea and the metropole led "to the conclusion that the foreign trade of Fernando Po [Bioko] and Rio Muni depends to a great extent on the rest of Spain, and that any drastic change in the present system of exchange would have serious consequences for both territories."[45] In the 1950s, the last full decade of colonialism, an outside observer also "ventured that Spanish Guinea will long remain fully dependent upon the metropole with little likelihood either of eventual independence or formal, full incorporation into the metropolitan system. Nationalist agitation is unlikely, for the native mass is isolated from liberal influences, it is inarticulate, and disorganized."[46] This assertion was tested by the wave of nationalism that submerged the island and the mainland in the late 1960s.

NOTES

1. Unesco, *The African Slave Trade from the Fifteenth to the Nineteenth Century. The General History of Africa, Studies and Documents*, Vol. 2 (Paris: Unesco, 1979), p. 212.

2. Pascal Bruckner, *The Tears of the White Man* (London: Collier Macmillan, 1986), p. xv.

3. Abelardo de Unzueta, *Islas del Golfo de Guinea* (Madrid: Instituto de Estudios Africanos, 1945), p. 174.

4. Robert Gard, "The Colonization and Decolonization of Equatorial Guinea" (Pasadena, Calif., unpublished manuscript, 1974), p. 24, quoting José Varella y Ulloa.

5. Paul Du Chaillu, *Explorations and Adventures in Equatorial Africa* (New York: Harper and Brothers, 1861; reprint, New York: Negro Universities Press, 1969), p. 49.

6. See David Northrup, "Nineteenth Century Patterns of Slavery and Economic Growth in Southeastern Nigeria," *International Journal of African Historical Studies* 12, 1 (1979), pp. 1–16.

7. Martin Lynn, "Change and Continuity in the British Palm Oil Trade with West Africa, 1830–55," *Journal of African History* 22 (1981), p. 340.

8. U.S. National Archives, Miscellaneous Letters of the Department of State, September 1–16, 1879, microcopy M-179, roll 545, William Thomson to Commodore R. W. Shufeldt, May 14, 1879.

9. Abelardo de Unzueta, *Geografía histórica de Fernando Póo* (Madrid: Instituto de Estudios Africanos, 1947), p. 163.

10. Billy Gene Hahs, "Spain and the Scramble for Africa: The 'Africanistas' and the Gulf of Guinea" (Ph.D. diss., University of New Mexico, 1980), p. 140, citing Ignacio Garcia Tudela, "Informe anual que el gobernador de Fernando Póo dirige al Exco. Sr. Ministro de Ultramar demostrado y encareciendo la convenciencia y la necesidad de abandonar las possessiones españolas del Africa occidental," MS. 1559, no. 17, Museo Naval, Madrid, Spain.

11. Teresa Pereira Rodriguez, "Las relaciones maritimo-comerciales entre Canarias y los territorios del Golfo de Guinea (1858–1930)," in *Las Canarias y Africa (Altibajos de una gravitación)*, ed. Victor Morales Lezcano (Las Palmas de Gran Canaria: Ediciones del Cabildo Insular, 1985), p. 70.

12. Max Liniger-Goumaz, *Connaître la Guinée Equatoriale* (Geneva: Editions des Peuples Noirs, 1986), p. 41.

13. Sociedad Fundadora de la Compañia Española de Colonización, *Memoria demostrativa de las ventajas y beneficios obtenibles de los territorios españoles del Golfo de Guinea* (Madrid: Imprenta de Fortanet, 1905), n.p.

14. See René Pélissier, "Uncertainty in Spanish Guinea," *Africa Report* 13, 3 (March 1968), p. 19.

15. Sanford Berman, "Spanish Guinea: Enclave Empire," *Phylon* 17 (December 1956), p. 360.

16. *Ibid.*, p. 359.

17. Georges Balandier, *The Sociology of Black Africa: Social Dynamics in Central Africa* (London: Andre Deutsch, 1970), p. 42.

18. *Ibid.*, p. 174.

19. Berman, "Spanish Guinea," p. 353.

20. Teresa Pereira Rodriquez, "Aspectos maritimo-comerciales del colonialismo español en el Golfo de Guinea," in *Segunda aula Canarias y el noroeste de Africa* (Las Palmas de Gran Canaria: Ediciones del Cabildo Insular, 1986), p. 251.

21. *Ibid.*

22. René Pélissier, "Fernando Poo: Un archipel hispano-guinéen," *Revue française d'études politiques africaines* 33 (September 1968), p. 98.

23. Manuel Góngora Echenique, *Angel Barrera y las posesiónes españolas del Golfo de Guinea* (Madrid: San Bernando, 1923), p. 116.

24. Gervase Clarence-Smith, "The Impact of the Spanish Civil War and the Second World War on Portuguese and Spanish Africa," *Journal of African History* 26 (1985), p. 324.

25. Román Perpiña Grau, *De colonización y economía en la Guinea española* (Madrid: Editorial Labor, SA, 1945), p. 162.

26. Berman, "Spanish Guinea," quoting *Africa* (Madrid, 1954), p. 313.

27. Gard, "Colonization," p. 92, citing *Anuario de estadistica y cadastro de la direccion de agricultura* (Dirección de Agricultura, 1944).

28. *Ibid.*, p. 152.

29. See Gerald Kleiss, "Network and Ethnicity in an Igbo Migrant Community" (Ph.D. diss., Michigan State University, 1975), p. 40; Mark Delancey, "Changes in Social Attitudes and Political Knowledge Among Migrants to Plantations in West Cameroon" (Ph.D. diss., University of Indiana, 1973), p. 103.

30. José Gutterrez-Sobral, "The Outlook at Fernando Po," *West Africa* (March 2, 1901), p. 334.

31. Anonymous memorandum, c. 1904, "Reorganization of the Administration with its Powers," John Holt Papers, John Holt and Company, Ltd., Box 10/6, Liverpool.

32. League of Nations, Secretariat, *Report of the Liberian Commision of Enquiry* (C.658.M272), June 1930, p. 36.

33. George Schuyler, "Wide 'Slavery' Persisting in Liberia, Post Reveals," New York *Evening Post*, June 29, 1931, p. 1.

34. S. O. Osoba, "The Phenomenon of Labour Migrations in the Era of British Colonial Rule: A Neglected Aspect of Nigeria's Social History," *Journal of the Historical Society of Nigeria* 4, 4 (1969), p. 520, citing Nigeria, *Annual Report of the Department of Labour 1954/55*, Lagos, p. 13.

35. Akinjide Osuntokun, "Nigeria–Fernando Po Relations from the Nineteenth Century to the Present" (Paper delivered at the Canadian African Studies Association Conference, Université de Sherbrooke, Quebec, April 26–May 3, 1977, p. 33.

36. R. J. Harrison Church et al., *Africa and the Islands*, 4th ed. (New York: John Wiley and Sons, Inc. 1977), p. 278.

37. Armin Kobel, "La République de Guinée Equatorialle, ses ressources potentielles et virtuelles. Possibilités de développement" (Ph.D. diss., Université de Neuchatel, 1976), p. 206, citing estimates of the Servicio Agronómico, Malabo, 1972.

38. Liniger-Goumaz, *Connaître*, p. 55.

39. "Update," *Africa Report* 14, 1 (January 1969), p. 26.

40. Gard, "Colonization," p. 871.

41. *Ibid.*, citing Commissary of the Economic and Social Development Plan, *Guinea Ecuatorial: Anexo al plan de desarrollo económico y social* (Madrid: Dirreción General de Maruccos y Colonias, 1964), p. 46.

42. Juan Velarde, "El Plan de Desarrollo Económico y Social de Fernando Póo y Rio Muni" *Archivos de Instituto de Estudios Africanos* 71 (1964), p. 14.

43. Berman, "Spanish Guinea," p. 359.

44. Gard, "Colonization," p. 406.

45. *Spain in Equatorial Africa* (Madrid: Spanish Information Service, 1964), p. 51.

46. Berman, "Spanish Guinea," p. 364.

3
Politics

Economic division in Equatorial Guinea plays an important part in explaining postindependence political machinations. Likewise, the fragility of inherited political institutions explains why they were so quickly swept away. The political milieu in the late 1980s was molded by economic and ethnic considerations that have little to do with the precepts passed on by a transient colonialism. The most important episode in the first twenty years of independence was the dictatorship of Francisco Macias Nguema. Unfortunately, the effects of his regime still haunt the republic.

DECOLONIZATION

It is significant to note that, in the 1960s when Portugal resolved to fight colonial wars, Spain resolved to grant independence to its sub-Saharan territories. Thus, Spain's policy was in line with the British and French policies of the early 1960s. The metropole, like Great Britain and France, hoped to retain some control after granting formal independence. Spain, however, envisioned the creation of a new government more democratic than the metropolitan regime. Between 1963 and 1968 the decolonization process was increasingly dominated by Spanish liberal and technocratic forces. Still, the outcome in Equatorial Guinea was not determined by the norms set out by the metropole, but by the dynamic forces of old and new African political movements. In retrospect, much of the constitution-making of the 1960s appears to have been an exercise in futility. The hastily conceived constitution was alien to the colonial power that retired from Equatorial Guinea in 1968. It was equally alien to the political realities that governed the lives of the majority of the new nation's citizens.

Spain's exit from Equatorial Guina can be traced to events that stirred Rio Muni in the late 1940s. The Fang and other groups on the mainland Muni were far more conditioned by ethnic revivalism and/

55

or African nationalism than their fellow Africans on Bioko. By the late 1940s the Alor Ayong movement in Gabon was having some effect in Spanish Guinea. The movement maintained the unity of three major Pahouin groups—Fang, Bulu, and Betsi—and asserted that they were descended from a common ancestor, Afri Kara. Alor Ayong urged the reunification of all the Pahouin clans. In 1947 leaders of the Fang met at Mitzic in the Woleu-Ntem region of Gabon. Delegates asked for the purification of customs and the election of a grand chief for the entire Pahouin group.

These developments doubtlessly had reverberations in Rio Muni. Leon Mba, the first president of Gabon and a participant in Alor Ayong, later claimed that he had early contact with dissidents in Rio Muni. In 1947 the Cruzada de Liberación Nacional (National Liberation Crusade) was organized there. The leader was Enrique Nvo, a schoolteacher from the Micromeseng region; his first followers were several seminary students. Members of the semisecret organization took a ritual blood oath and were christened with seawater, which was thought to increase fraternal bonds. A group that later came to be known as the "death quintet"— Esteban and José Nsue, Justino Mba, Agustin Eñeso, and Frederico Ngoma Nandango—was the core. Secondary-school students provided new recruits. Also, older members, like Atanasio Ndongo, Salomé Jones, Luis Maho, and Acacio Mañe became part of the leadership cadre shortly thereafter. Jones had extensive political contacts in anglophone West Africa, and Mañe, a coffee grower, provided funds.

The Spanish soon acted to suppress the Cruzada Nacional. In 1952 Atanasio Ndongo and Enrique Gori Molubela, a seminary student on Bioko, organized a strike of seminarians. Spain ordered the movement dissolved and Ndongo and Gori were expelled from school. Ndongo left Spanish Guinea in 1951; first he lived in Yaoundé and later in Gabon, where he had contact with nationalist circles among the Gabonese Fang. Eventually he returned to Cameroon and formed a political party. In late 1961 he toured the Nsork, Ebebiyin, and Mongomo districts on the border of Rio Muni and rallied crowds to his brand of nationalism. In Libreville, Gabon, in March 1962 he took the helm of the Movimiento Nacional de Liberación de Guinea Ecuatorial (MONALIGE). After forming this group, Ndongo made a tour of African, European, and Asian countries.

Maho continued a separate nationalist tradition in Bioko. In September 1961 he proclaimed himself the leader of the Cruzada Nacional de Liberación de Guinea Ecuatorial (CNLGE). In 1962 he hurriedly left Bioko and made his way to Gabon, where he made contact with Ndongo's group. The collaboration was short-lived; in October 1962 Maho moved to Douala, Cameroon, and gave up the Cruzada Nacional. He switched to the Movimiento Pro-Independencia de Guinea Ecuatorial (MPIGE), which was headed by Pastor Torao, the mayor of Santiago de Baney on Bioko.

Yet another pro-independence group, the Idea Popular de Guinea Ecuatorial (IPGE), had been formed in 1959. This party petitioned the United Nations and complained about Spain's administration of the colony. In the aftermath of the petition, several members of the IPGE were arrested. One of the chief spokesmen of the movement, José Perea Epota, favored union with Cameroon.

In February 1963 Ndongo and Maho met with Epota in Ambam, Cameroon, and established the Coordinating Bureau of Guinean Movements. Maho was president, Perea Epota was vice-president, and Ndongo was secretary-general. Unfortunately, because the group was torn by personal disputes and wrangling, it was disbanded in June 1963. A stumbling block had been IPGE advocacy of union with Cameroon. Later the position of merger with Cameroon produced splits within the IPGE itself.

By the 1960s Spain's position as a colonial power in sub-Saharan Africa was not popular. There was talk of annexation by one of Equatorial Guinea's newly independent neighbors. Gabon surrounded Rio Muni on three sides, and the Fang on both sides of the border had many commonalities. Plans for unification were intermittently discussed. A serious possibility of annexation also came from Nigeria.

In the face of these threats, Spanish policy vacillated. Madrid reversed its policy of calling the African territories overseas provinces. In August 1963 it announced its intention to give the African provinces autonomy and drew up the Basic Law on the Autonomy of Equatorial Guinea. On December 20, 1963, Spain's Equatorial Region was officially renamed Equatorial Guinea.

The plan to give the colony greater autonomy was submitted to the voters in Equatorial Guinea. Most nationalists campaigned for a yes vote: Jaime Nseng, former secretary-general of IPGE and now leader of the Movimiento Nacional de Unión (MNU); Bonifacio Ondo Edu of the newly founded Movimiento de Unión Popular de Liberación de la Guinea Ecuatorial (MUPGE) and a large part of the MONALIGE. There were 57,244 votes cast for the new autonomy statute and 35,365 votes cast against it. On Bioko the vote was heavily against the statute; the old island elites did not want to be deluged by mainlanders in a new system of government. In addition, the island followers of MONALIGE opposed the law because they maintained that too much power remained in the hands of the Spanish general-commissioner.

In Rio Muni the situation was far more complicated. Most all of the nationalists wanted independence but personal and ethnic rivalries skewed the picture. The Ndowe and other peoples of the coastal region voted for the change because it seemed to promise to maintain their position vis-à-vis the much more numerous Fang. In the interior districts inhabited by Fang, there were divisions in voting patterns between the Ntumu Fang north of the Mbini (Benito) and the Okak Fang south of it. The former generally favored union with Cameroon. Thus the IPGE was successful in getting a no vote in the north. In the south Ondo Edu and the MUPGE campaigned for a yes vote and continued—perhaps temporary—association with Spain. This pattern was broken by the voting in the corner districts of Mongomo and Ebebiyin, areas north of the Mbini, where the MNU campaigned and won a yes vote.[2]

The 1963 autonomy statute, the Basic Law, established the Council of Government, a court system, and the General Assembly. The latter chose the eight counsellors of the governing council, although its choices had to be approved by the Spanish head of state. As a result of elections held in March 1964, 58 persons were elected to municipal and village councils by family heads: Ten were chosen by cultural, economic, and professional organizations; 10 were elected by cooperatives; and 38 were directly appointed. The 116 council members met in April to vote for the members of provincial deputations. The Provincial Deputation of Rio Muni chose Federico Ngomo Nandongo as its president. The other segment of the colony chose Enrique Gori Molubela. The two provincial deputations met jointly in May to select the 8 members of the Council of Government, 4 from Rio Muni and 4 from Bioko. Subsequently, the Spanish government chose Bonifacio Ondo as the president of the council.

From the inauguration of the autonomous regime in 1964 onward, nationalist movements were recognized legally as political parties. Several parties emerged; none was strongly differentiated by ideology and membership was shifting. A French political scientist observed that elections "were dominated by strong personalities and ethnic considerations rather than party politics."[3] By 1968 there were five contending groupings, all but two advocating independence and strong union between Rio Muni and Bioko: MUNGE (Movimiento de Unión de Guinea Ecuatorial), MONALIGE, IPGE, Unión Democrática Fernandina, and Unión Bubi.

Unión Democrática Fernandina and Union Bubi, both based on Bioko, had separatist platforms. Whereas the latter favored total separation from Rio Muni to protect Bubi interests, the former sought to protect powerful Spanish interests on Bioko through a federal solution that would secure the island from mainland domination.

Of the nationalist groups, MUNGE appeared the most willing to compromise and received the support of the Spanish authorities. It had been founded in Bata just before the autonomy referendum of 1963.

MUNGE's founders aimed to unite all existing political groups into a single unified movement similar to Franco's Falange. MUNGE opposed union with any foreign power as interference in the negotiations between Equatorial Guinea and Spain, a stance directed at Cameroon. It also called for the political union of Bioko and Rio Muni and a regime based on "African Christian socialism." Initially, the group gained an impressive following. Most of the nationalists who were not in exile responded with enthusiasm. The group included almost all political tendencies within the colony: Salomé Jones; Bonifacio Ondo Edu; Francisco Dougan (a Fernandino with a large following); and Enrique Gori Molubela.

After the autonomous government was installed, MUNGE proceeded to co-opt independent officeholders, including Bonifacio Ondo Edu. By the end of 1964, MUNGE had swallowed up almost all of the significant African officeholders. Yet, because of its perceived closeness to the colonial regime, many of the rank and file became disenchanted. The other main independence parties increasingly came to the forefront. MONALIGE proved less docile than MUNGE, although it acted within the guidelines set down by Madrid. It accepted autonomy as the first step toward independence, but had strong reservations about the power remaining in the hands of the Spanish commissioner-general in the "autonomous state." The party was handicapped at the time of the 1964 elections by the absence of Ndongo.

Although many MONALIGE partisans went over to MUNGE in the period immediately following the 1963 referendum, they gradually returned. As MUNGE increasingly showed its unwillingness or inability to quicken the pace of decolonization, MONALIGE gained support. In October 1966 Ndongo returned from exile, bringing with him his future wife, the widow of the assassinated leader of the radical Union des Populations du Cameroun. By 1968 a rift had occurred in the party. Francisco Macias Nguema, vice-president of the Consejo del Gobierno Autonomo and *consejero de obras publicas* (director of public works) denounced Ndongo's leadership and claimed to be the party leader.

To the left of MONALIGE was the Idea Popular de la Guinea Ecuatorial (IPGE). This group included a few Marxists, as well as nationalists. IPGE Secretary-General Jesus Mba Ovono was operating from exile when autonomy was granted. In October 1964 the exiled leaders of MONALIGE and IPGE met in Accra and announced they had joined forces in a new organization: the Frente Nacional de Liberación de Guinea Ecuatorial (FRENAPO). However, this union proved extremely fragile and the alliance very soon fell apart. IPGE itself fell victim to its own factionalism. Eventually an exile group headed by Jesus Mba Ovono established itself at Brazzaville while a legally recognized group, headed by Clemente Ateba, operated within Equatorial Guinea.

On Bioko politics took a far more conservative turn. Many Bubi agriculturalists and functionaries were tied to the presence of oligopolistic colonialism. Superficially, the cleavage appeared ethnic (Fang versus Bubi). More fundamentally, it was a manifestation of the different rates at which European capital had penetrated the two segments of the colony. On the eve of independence, one observer said: "The Bubi, who are indirectly benefitting from the flight of Spanish investment from Rio Muni, are disenchanted with continental nationalism. They are now convinced that independence in union with Rio Muni, and the loss of Spanish protection, would open their island to economic and political plunder from a flood of Fang."[4] In August 1964 a formal separatist campaign began with a meeting of the presidents of all the village councils. The meeting produced a petition asking for administrative separation from Rio Muni, a call that was repeated several times in the following years; in March 1968, Bubi from all sections of the island held a huge meeting at Basupu del Oeste and drew up a petition asking for separation.

Some segments of the Spanish government favored independence for Madrid's small sub-Saharan territories. The Foreign Ministry became convinced that the independence of Equatorial Guinea could be traded for anticolonialist support in the United Nations on the status of Gibraltar. The ministry's viewpoint was particularly important because, by the mid-1960s, the organization was taking a hard look at Spanish colonialism. In late 1965 the U.N. General Assembly met and heard Bonifacio Ondo and Atanasio Ndongo. The latter bitterly attacked the colonial regime and alleged that MUNGE was a puppet of Spain. In December, in the General Assembly, 103 members voted for Resolution 2067, which demanded the setting up of a date for independence. In April 1966 all African government functionaries went on strike; on April 29 in Bata the members of the colonial General Assembly voted unanimously to boycott the functions of the Council of Government. They asked the commissioner-general to hold an election for a new council. As part of their demand, they asked that only Africans be allowed to vote; this was intended to dilute the power of the resident Spanish population.

A visit by the U.N. Committee of Twenty-Four on Decolonization in August 1966 strengthened the hand of the unionist parties. The subcommittee insisted on talking to MONALIGE and IPGE leaders, as well as those of MUNGE and the Bubi separatist group. MUNGE agreed to give Spain two more years to decolonize. However Francisco Macias Nguema, leader of one splinter of MONALIGE, was insistent and demanded immediate independence.

In the mid-1960s Macias Nguema, the future president of Equatorial Guinea, was an unknown quantity. He appeared both a fervent anti-imperialist and ardent admirer of Francisco Franco. Born in 1924 (or 1920), the future president had an unspectacular, if somewhat opportunistic, career as a colonial functionary. Although he was to become the first president of Equatorial Guinea, he may have been born in Gabon.[5] He received a basic education at Catholic mission schools. In 1940, 1941, and 1942, he failed entrance examinations for the colonial civil service. In 1943 he did gain employment in the *sub-gobierno* of Bata with the Forest Service and Public Works Department. A year later, with the strong recommendation of his Spanish superiors, he took the examination that led to his attainment of *emancipado* status. The future president, once known as Mez-m Ngueme, Hispanicized his name to Macias Nguema Biyogo Negue Ndong. He subsequently became an auxiliary administrator, first in the Rio Benito district and later in Bata with the public works department, where he remained until 1962. Macias Nguema then spent a year in Cameroon after suffering a long illness. On his return to Rio Muni, he became assistant interpreter in the native tribunal in Mongomo, where he reputedly used his position to curry favor with the authorities and to extort favors from litigants. Macias Nguema's efforts were eventually rewarded by the Order of Africa and the *Merito Civil*, as well as by the mayoralty of Mongomo, where he served from 1963 to 1968.

In the 1960s Macias Nguema made two trips to Spain, one to render homage to Franco in the name of Guinean functionaries. When he did enter politics, he displayed a chameleonlike ability to shift party loyalties. In 1963 he joined the IPGE. In 1964 he moved to the MUNGE in the period of co-optation following the granting of autonomy. He subsequently shifted allegiance and joined the MONALIGE, where his presence was highly disruptive; he quarreled with Ndongo and split the party.

In August 1966 a U.N. subcommittee visited Equatorial Guinea and reported that the inhabitants wanted and should have independence no later than 1968. Later, in December 1967, the U.N. General Assembly called upon Spain to promise that Equatorial Guinea would accede "to independence as a single political and territorial entity not later than July 1968."[6]

Spain agreed to a conference on the colony's legal status to be held in late 1967. Its first phase began in Madrid in November 1967. There were 42 active delegates, 27 of whom favored unified independence. Fifteen members favored some kind of separation between the island and the mainland. The spokesman for Bioko interests, Enrique Gori, strongly asserted the island's right to self-determination and argued that union imposed an economic burden. On the other hand, some of the delegates favoring unified independence asked for the immediate proclamation of a provisional government. Leaders of various unionist factions formed a "conjoint secretariat" (*secretariado conjunto*) in which Francisco Macias Nguema emerged as the main spokesman.

After nine inconclusive meetings, the Spanish Foreign Ministry recessed the colonial conference sessions until April 1968. The second phase of the conference, which had more than 40 delegates, lasted from April 17 to June 17, 1968, and was characterized by the last-ditch efforts of Bioko's community to obtain either separate independence or no independence at all. After much discussion, a unitary constitution was announced. It established a presidential system with separate provincial governments for Bioko and Rio Muni. On July 9, Spain announced that the vote for the constitution in Equatorial Guinea would take place on August 11 and the result would be taken as a yes or no on independence.

Some African participants were wary of plans for both the use of Spanish currency in independent Equatorial Guinea and the continued presence of Spanish troops. Macias Nguema was a prominent critic of the proposed terms of independence. On July 17, he and a group of colleagues spoke before a U.N. committee in New York. They said that the proposed constitution and electoral law were specifically designed to allow Spain to exercise neocolonial control over its former territories.

On July 19, the U.N. committee adopted a resolution that regretted the cleavages in Equatorial Guinean opinion and promised that a U.N. presence would supervise the constitutional referendum. Just before the referendum, there were demonstrations in Malabo against proposed independence and about one-third of the eligible voters on Bioko abstained. Nonetheless, the referendum took place as scheduled on August 1. Four days later the official results were announced: Of 115,885 ballots cast, 72, 458 favored independence under the proposed constitution and 41, 197 were opposed (2,198 ballots were declared invalid).

Moving rapidly toward decolonization before the end of 1968, Spain announced on August 21 that presidential and general elections would be held on September 22 and independence would be granted on October 12. As in other instances of decolonization, intense competition arose over the specific transfer of power. In a first round of voting on September 22, no candidate received the required absolute majority: Macias Nguema, head of a vague group of dissidents from other parties, received 36,716 votes; Ondo Edu of MUNGE received 31,941 votes; Ndongo Miyone of the main wing of MONALIGE received 18,232 votes; and Edmundo Bosio Dioco of the Unión Bubi received 4,795 votes. In the September voting for the 35 members of the National Assembly, 8 seats went to Macias Nguema's IPGE-MUNGE-MONALIGE coalition (Grupo Macias), 10 to Ndongo's MONALIGE, 10 to MUNGE, and 7 to the Unión Bubi.

Election scene at a polling station in Rio Muni, 1968. (Photo courtesy the United Nations)

A week later, second-round elections were held. This time Macias Nguema won 47,400 votes, while Ondo Edu received 24,000.

The triumph of Macias Nguema may be attributed to several factors. Perhaps the most important factor was the rivalry between his opponents. Ondu Edu and Ndongo were better known but bitterly opposed to each other. They had failed to form an electoral alliance. After much backstage maneuvering, Ndongo and Edmundo Bosio Dioco, who had the support of some MUNGE and MONALIGE dissidents, threw their weight behind Macias Nguema.

THE REGIME OF MACIAS NGUEMA

On October 13, the Spanish minister of information, as representative of Generalissimo Franco, devolved sovereign power on the government of Macias Nguema, saying that it was the "result of a peaceful, friendly, and constructive development."[7] The new president emphasized the need for continued collaboration and expressed his gratitude, although with some reservations. Macias Nguema's combination of militancy and opportunism gave contradictory indications of his future intentions. Political offices were parceled out to rivals. For example, Atanasio Ndongo became foreign minister. The inclusion of a man Macias Nguema had

64

POLITICS

strongly criticized seemed to promise a regime based on diversity of opinion and reconciliation.

Such hopes did not survive the first six months of independence. The fragility of the personal and political alignments should have been evident. The overwhelming balance of power lay with Macias Nguema. The council of ministers, over which Macias Nguema presided, was dominated by his followers, particularly his cousin and minister of the interior, Masie Ntutumu. The holder of potential checks on the presidency, Moses Mba Ada, president of the Consejo de la Republica, was also one of Macias Nguema's partisans.

Barely four months after independence a grave crisis arose that signaled the total removal of all obstacles in Macias Nguema's way. To counter the Bubi-Spanish alliance, the president brought some 7,000 Fang to Malabo. He attacked the former colonial power in speeches and his new government insisted on reducing symbols of Spanish presence. The president's youth movement demonstrated in Bata and burned a Spanish flag. Later, during a riot in Mbini, a Spanish foreman was killed and the Spanish ambassador, Juan Duran Loriga, mobilized the 260-man Spanish Guardia Civil (which had been permitted to stay in Equatorial Guinea under the terms of independence). The Guardia occupied the Bata and Malabo airports and the telecommunications centers. At the same time, Spanish vessels arrived at Bata, ostensibly to protect Spanish nationals. Macias Nguema charged that the ambassador had ignored the terms of independence, which allowed the guardia to be used only with the mutual consent of the two countries. Ambassador Duran Loriga and the Spanish consul general in Bata were declared personae non grata; the Spanish government recalled them on March 2, 1969.

Equatorial Guinea declared a state of emergency, imposed a dusk-to-dawn curfew, demanded the evacuation of Spanish forces, and urgently appealed to the United Nations for 150 peace-keeping troops. Spaniards who feared maltreatment began an exodus that continued until the end of March. By the middle of March 400 Spaniards had left. At the end of the month most of the evacuation was complete. Only 80 Spaniards chose to remain in a country that had contained over 11,000 Europeans in 1960. During this crisis Spain sent a special envoy as provisional ambassador. This envoy was unable to start immediate negotiations with Macias Nguema, but did order the Guardia Civil to return to barracks, where they remained until their evacuation.

Foreign Minister Atanasio Ndongo and Saturino Ibongo, the Equatorial Guinean delegate to the United Nations, were asked by Spanish officials to stop inflammatory broadcasts from Equatorial Guinea. Both men visited Spain on their way back from an Organization of African

Unity (OAU) meeting in Addis Ababa in February 1969. On their return to Equatorial Guinea, the two attempted a coup. After meeting with Macias Nguema in Bata on March 4, Ndongo proclaimed himself president but was soon routed. The followers of the ill-fated putsch fled to the forest while the president ordered the arrest of eleven political leaders. Ndongo and Ibongo died in detention and an unknown number of suspects were liquidated. Ondo Edu, who had not participated in the plot, was arrested largely because he was viewed as a future opponent.

The abortive coup became the rationale for the subsequent ruthless liquidation of both real and potential opposition. In 1975 the Equatorial Guinean representative at the United Nations cited the events of the postindependence year as those with the most significance for Equatorial Guinea: "The thwarted coup of the former Minister of Foreign Affairs, Atanasio Ndongo Miyone, who was bought by the Spanish colonialists for a price of 50 million pesetas (about U.S.$892,858) and the expulsion of all the Spanish colonial military forces . . . together with colonial civilians marked the end of colonialism in Equatorial Guinea." Ndongo's attempted seizure of power fueled the subsequent paranoia that characterized national policy. According to the government, the coup was led by "men who confused ambition with patriotism, persons who expected something distinct from what they saw on October 12 [Independence Day], groups of people with their consciences reduced through lack of love for their people, carried away with selfishness in order to drive Equatorial Guinea to chaos, selling our people into the hands of a foreign people."[8]

From 1969 onward, the twin props of Macias Nguema's regime were the Juventud en Marcha con Macias (Youth on the March with Macias) and the National Guard. These two agencies ensured, through expropriation of private property, intimidation, and political murder, that the powers of the head of state were not challenged. From 1969 to 1979, thousands fled their homeland and sought refuge in Cameroon, Gabon, Nigeria, Europe, and America. Estimates as to the number of people killed vary greatly. According to some, about one-third of the approximately 300,000 people of Equatorial Guinea were killed and another 50,000 fled abroad.[9] In 1974 it was estimated that there were 60,000 refugees in Gabon, 30,000 in Cameroon, 15,000 in Nigeria, and at least 6,000 in Spain.[10] Another 1,000 were in Europe and America.

Critics of the regime charge that the dictatorship had a disastrous effect on the overall population. It has been estimated that the population should have amounted to more than 400,000 at the end of the 1970s. However, according to opponents of the former regime, the population was no more than 250,000. In 1968 the rate of increase was 1.9 percent per annum. The mortality rate was 7.8 percent (against 27 percent for tropical Africa). According to one source, mortality rose to 21 percent per annum in 1978.[11]

Repression in Equatorial Guinea may have been perceived to have an integrative function. Terror intimidated the recalcitrant insular population and, at the same time, promised rewards to a politically dominant group. The president used the instruments of the modern state to mold a new unit appealing to Fang nationalism and disrupting the structures and classes bequeathed by the colonial state (i.e., the metropolitan and resident European capitalists, the black bureaucracy, wealthy black cocoa and coffee growers, and the Igbo migrant labor population).

Macias Nguema's victory was the conquest of one member of one segment of an ethnic group. The dictatorship was not so much a triumph of Fang nationalism as it was a triumph of the president's hometown coterie. His power base was in the Mongomo district, especially among members of the Esangui clan. Although the regime attempted to appear populist, its limited base among a segment of the Fang made it highly unrepresentative. The Esangui are among the least numerous of fifty ethnic divisions on the mainland. In 1952, 10 percent of the villages in the Mongomo district were peopled by Esangui and only 11.3 percent of the population belonged to the clan. The district contained 12,039 inhabitants in 1963, 5.3 percent of the total population of Equatorial Guinea and 6.6 percent of Rio Muni's population. In the same year, the district's voters (3,441) represented only 2.7 percent of the national electorate. At independence, some 4,000 Esangui represented barely 1.5 percent of the total population. Significantly, in the final presidential election of 1968, the area voted against Macias Nguema.[12]

The president circumvented or daunted traditional rulers and, in the name of African authenticity, replaced them with members of his own family. Ela Nguema, the president's nephew, served as his aide-de-camp. A cousin, Masie Ntutumu, at one point served as minister of interior. Eyegue Ntutumu, a kinsman, was appointed governor of Rio Muni in 1971. After the execution of Ndongo, Nguema Esono, another cousin, became foreign minister; in 1976 he took over as vice-president. At one point, Oyono Ayingono, another nephew, held several posts simultaneously: finance minister, industry and commerce minister, information director, and secretary of state to the presidency. In addition, he was chief of protocol and commissioner of state enterprises. Feliciano Oyono, also a cousin, was permanent secretary of Nguema's political party and Teodoro Obiang Nguema, a nephew, was head of the National Guard.

Macias Nguema's youth movement, the Juventud en Marcha con Macias, was composed of marginally employed men and in some ways was comparable to Haiti's Tonton Macoute. The Juventud proved to be

Portrait medallion of Macias Nguema. (Photo courtesy Max Liniger-Goumaz)

one of the most important props of the regime. The juventud was the most visible agent of the government and could sustain itself through the expropriation of goods belonging to persons without official protection. Given the great disparity between the two halves of the republic, the movement of unemployed youths to richer areas was both a safety-valve and a way of enforcing the will of the president.

In 1970, a year after the abortive coup, the government abolished rival parties and substituted the Partido Unico Nacional (Sole National Party). Later, the party was rechristened the Partido Unico Nacional de Trabajadores (PUNT), Sole National Workers' Party. In 1972 Macias Nguema became president-for-life and in July of the following year a new constitution was approved by PUNT and ratified by referendum. The People's National Assembly was to be appointed by PUNT and could be removed by it at any time. All judges and public prosecutors were nominated by the president. Significantly, those accused of subversion had no rights. Macias Nguema was not bound by constitutional considerations. For example, the vice-presidency should have been occupied by an islander. But after Vice-President Edmundo Bosio Dioco fell from grace, the Bubi was replaced by a Fang from the president's hometown.

Under the dictatorship, there was greater integration in the administration of the island and the province. Malabo remained the official capital, but the locus of power shifted to Mongomo where Macias Nguema spent most of his time. Each province was given a "gorci" (*gubernador civil*) and each district was governed by a *delegado gubernativo*.

The civilian authority was hemmed in by that of two military officials, one from the National Guard and one from the militia, each of whom commanded detachments of troops. The civilian *delegado* oversaw a subordinate committee composed of the local president of PUNT, the president of the women's section of PUNT, and the president of the youth wing. In theory, the structure was under the control of the party and the military, but responsible to the civilian administration. In reality, the security apparatus was largely autonomous. At the grassroots level, Macias Nguema's followers were increasingly dominating traditional structures; he selected militants to replace traditional chiefs. *Comités de base*, composed of the local president of the women's section and the president of the local Juventud, existed in every village. One of the committee's chief functions was surveillance and control of movement. Movement was also regulated by checkpoints throughout the country, which were manned by counterbalanced units of the National Guard and the militia.

Political opponents were among the first to feel the vengeance of the regime. In November 1968 a chief political opponent, Bonifacio Ondo Edu, was starved and later given the coup de grace in prison. Many others simply disappeared. In 1970 the majority of the Bubi politicians were eliminated. In June 1974 no less than 118 prisoners were tortured and killed in Bata's central prison. The following year, former Vice-President Edmundo Bosio Dioco "committed suicide" in mysterious circumstances after writing a letter pledging loyalty to "Comrade President-for-Life Francisco Macias." In 1976 exiled students occupied the Equatorial Guinean embassy in Madrid and, as a result, there were reprisal killings of their relatives in the republic. An outbreak of cholera on the island of Annobón (Palagú) was allowed to run its course because the president apparently doubted the political reliability of its inhabitants. Of the 46 politicians who attended a Madrid constitutional conference in 1967–1968, not more than 10 were alive at the time of Macias Nguema's fall. Outside of a few who had died of natural causes, the vast majority had been put to death. The same holds true for the first legislative assembly; over two-thirds of its members either died violently or disappeared.

Terror

Those accused of offenses against the security of the state had no specific rights. The National Guard and the Juventud followed no law except the will of the president. Macias Nguema appointed all judges and public prosecutors. Rigid postal censorship, the abolition of passports, and the eventual destruction of canoe traffic from Bioko reinforced the citizens' isolation. The very vagueness of being charged a *descontento* (often a capital charge) abetted collaboration with the government. State violence was largely extraconstitutional. Because the president-for-life was the source of all rewards and sanctions and because no act committed in his name could be called into question, the court system was bypassed in favor of a direct system of punishments administered by the security apparatus. Trials were rare; suspected miscreants were imprisoned without due process.

The president-for-life headed the security apparatus and controlled its chain of command. This ran through the national director-general of security and the provincial governors down to the district heads of the militia (district representatives of the Juventud). At the village level there were local *jefes de seguridad* (security chiefs), who often relied on denunciations brought in by village youngsters. Suspects were arrested and interrogated, often under torture. The director-general of security and other officials reviewed the accused's conduct and then placed the case before the president-for-life. There was no trial, no definite sentence, and no defense. It is important to note that in Equatorial Guinea political prisoners fared worse than common criminals. The aim was to destroy political prisoners either physically or morally "since their death was of no account." No jailer "was accountable for dead prisoners and quite often guards were ordered to kill them, for which they were rewarded."[13]

Death sentences, a foregone conclusion if the accusation came from Macias, were carried out by public executions with obligatory cheering attendance from all the population within reach. Methods have varied. In 1969, the victims "were unskillfully hanged . . . to the strains of Mary Hopkins singing 'Those Were the Days' over the loudspeaker system." [*Financial Times*, February 17, 1970] The background music was not used later, possibly because of breakdowns in the electrical system. On Christmas Eve of the same year, Malabo prisoners were publicly shot, but later executions have usually not involved firearms. Instead prisoners have been beheaded, with their heads left to rot on poles, strangled or beaten to death. In the case of the Director of the National Bank in Malabo, Mr. Buendy, his hair and eyes were burnt before he was killed in a spectacular event in Bata in 1976. Subsequently, his village was destroyed and the remaining villagers beaten to death.

Other wholesale killings have occurred in villages connected to some offender and, at times, when the actual prison was too small for expediency. This occurred in Bata, in 1974, when thirty-six prisoners were taken out of jail and ordered to dig a ditch. They were then forced to move into it and earth was filled in, leaving only their heads above ground. The next day witnesses saw that all but two were dead with their eyes missing and their faces partly eaten by insects.[14]

Insofar as it had any ideology, the regime was reasonably consistent only in its militant anticolonialism. Initially, it avoided declarations of socialism and instead proclaimed a vague African humanism. It specifically condemned Marxism. Macias Nguema's own political thought was a curious blend of anticolonialist rhetoric and *Machtpolitik*. Although intensely hostile to the former metropole, he had on occasion paid homage to Franco and called Adolf Hitler "the saviour of Africa."[15] He also admired a diverse assortment of African leaders: Idi Amin, Haile Selassie, Mobutu Sese Seko, Marien Ngoubi, and Jean-Bedel Bokassa. In the early days of his regime, Macias Nguema did not question the role capitalism was to play in his country. In 1971 he announced that his people should do all in their power to encourage a climate favorable to foreign investment.

Relations with Foreign Countries

The fear of violent overthrow by elements opposed to the regime was a constant preoccupation. In 1970, following the Portuguese attack on Guinea [Conakry], there were outbursts against Portuguese living in and around Malabo. According to Portuguese refugees, the president organized an anti-Portuguese rally and members of the Juventud attacked Portuguese settlers wherever they could find them. A real invasion attempt was planned in 1972 by a band of mercenaries under the leadership of Frederick Forsyth, British journalist and author of *The Dogs of War*. Forsyth and Alex Gay, a Scot with mercenary experience in Zaire and Biafra, plotted to use 50 former Biafran soldiers and 12 European soldiers of fortune in a coup. The plotters purchased a ship and arms but were stopped by Spanish authorities in the Canaries. Rumors said that Admiral Carrero Blanco or the former head of Biafra, Chukwemeka Ojukwu, was behind the effort. Either way, the episode increased the vigilance of the already paranoid Macias Nguema.

In 1973 a Nigerian formerly employed by the U.N. mission on Bioko reported that four foreign employees of the U.N. Development Program were deported for allegedly taking part in a plot against the president. Subsequently, the development program was closed. The previous year, in an interview with the Cuban Press Agency, Macias Nguema accused colonialists of attempting to divide the national patrimony. In his opinion, Equatorial Guinea's strategic position and the possible presence of oil made the NATO powers extremely interested in the area.

His relations with his closest neighbors were mercurial. Independence destroyed the anticolonialist rationale for annexation and presented new problems. Whereas, as late as May 1965, the Nigerian federal legislature could hear a strong motion "to enter into negotiations with the Spanish authorities with a view to acquiring the island of Fernando Poo," by 1968 the island's independence was a respected fait accompli.[16] During the Biafran War, the presence of thousands of Igbo workers presented the government with hard decisions. The new republic became a possible ally to either of the participants in the Nigerian Civil War. The Biafran regime was anxious to solicit Nguema's friendship. At the time of Equatorial Guinea's independence celebrations, Colonel Ojukwu sent a congratulatory message thanking the new president for aiding the "traumatized people of Biafra." The newly achieved independence was hailed as a "great victory" and the Guineanos were reminded that "Biafrans" had resided among them for decades.[17]

Ojukwu's friendly gestures were rebuffed. Macias Nguema was perhaps resentful of Spain's support of the break-away regime and the rumors that Ojukwu might flee to Bioko and continue his resistance. Equatorial Guinea drew closer to the Lagos government. At the end of 1969 the little republic announced the suspension of nightly Red Cross relief flights. Jesus Alfonso Oyono, minister of public works, announced that as a member of the Organization of African Unity his state would not aid Biafra. Macias Nguema addressed the Nigerian community in Malabo and said that he would negotiate an agreement on daytime relief flights. The existing relief arrangement had been negotiated by Spain and was "not in accord with the policy of Equatorial Guinea."[18] The dictator's action may have been more than an expression of solidarity with the beleaguered Nigerian government. Antonio Garcia-Trevijano, Nguema's European adviser, warned that Biafra was being sponsored by a number of European banks, including the Rothchilds. The continuation of the Biafran struggle might involve the seizure of Bioko as part of an Igbo redoubt.

On Equatorial Guinea's first anniversary, a telecommunications link with Lagos was established. General Jakubu Gowon described this as a symbol of their friendly relations. Subsequently, Biafran radio attacked the Equatorial Guineans for alleged brutalities against migrants who refused to support the Federal Nigerian government. In August 1970, after the collapse of the secessionist movement, Macias Nguema visited Nigeria and reemphasized the cordial relations existing between the two governments. He labeled rumors of annexation by Nigeria as untrue and added that postindependence contact between the two peoples would remove the "burdens" imposed by the ex-colonialists. Relations did not improve, however. The continuing abuse of Nigerian laborers was a sore point. By 1976 it appeared that Nigeria might intervene directly to protect the rights of its nationals. President Murtala Muhammad's assassination diverted Lagos's attention from its troublesome neighbor.

Relations with Gabon also deteriorated. In the early 1970s a dispute with Gabon over the islets of Cocotiers and Mbanie brought the two states to the brink of war. There were divergent styles of government (erratic anticolonialism versus Francophilia) and the problem of smuggling on the Rio Muni–Gabon border. In August 1972 Gabon extended its territorial waters to 100 nautical miles, which promised to deny Equatorial Guinea potential oil revenues.

Furthermore, also in August 1972, Nguema declined to visit Gabon's independence-day festivities. Shortly thereafter, Equatorial Guinean civil guards were sent to occupy Mbanie and Cocotiers. This preemptive move provoked Gabon to send police to protect Gabonese fishermen. On September 8, President Omar Bongo informed the U.N. secretary-general that the islets had been attacked by the forces of Equatorial Guinea. Equatorial Guinea charged that Gabon had made naval demonstrations at the Rio Muni estuary and destroyed some of its shipping. Gabon's diplomats were ejected from Bata and an anti-Gabonese demonstration was organized in the streets.

The United Nations chose to leave the dispute in African hands. A conciliation conference was organized under the leadership of Presidents Mobutu (Zaire) and Ngouabi (Congo People's Republic) and a mediation conference was convened in Kinshasa with both Bongo and Macias Nguema present. It was quickly agreed to appoint a commission to recommend a solution. This seemingly quick resolution, however, was stillborn; Gabon refused to withdraw its forces from one of the disputed islets. It also accused Equatorial Guinea of creating unrest among the Fang of the Woleu-Ntem area by stressing that Equatorial Guinea's dispute was with Bongo rather than the Fang. Later, both parties consented to "neutralization" of the disputed portion of the Bay of Corisco and to an OAU commission to define the maritime boundaries of the two states. In a postcrisis mood, Bongo and Macias Nguema also discussed strengthening relations between their states. Less than a year later, Macias Nguema paid a three-day visit to his former antagonist, supposedly signaling the subsidence of the "sandbank" crisis. The OAU mediation of the Gabon–Equatorial Guinea dispute was seen by some observers as a triumph of inter-African diplomacy. In 1974 dispute over offshore territories flared up again, this time exacerbated by the continuing influx of Equatoguinean refugees into Gabon. In 1977 Bongo visited Spain and is thought to have asked for Madrid's diplomatic support against its former colony.

Equatorial Guinea's relations with its other neighbor, Cameroon, were less dramatic than those with either Nigeria or Gabon. Cameroon did not have the large migrant colony in Equatorial Guinea that Nigeria had, nor did it share the same close ethnic bonds. However, during the

period before Equatorial Guinea's independence, many Guineano politicians sought refuge in Douala and elsewhere. Yaoundé made several
friendly overtures toward its tiny neighbor during its first year of
independence. Equatorial Guinean Vice-President Edmundo Bosio Dioco
visited Cameroon in August 1969 and laid plans for a treaty of friendship
and economic cooperation. The completed agreement was signed by
Macias Nguema on a state visit to Cameroon in January 1970. The
Equatorial Guinean president spoke of his foreign policy as "cooperation
with African States and in particular with Cameroon."[19] In 1971 President
Ahmadou Ahidjo became the first head of state to visit Equatorial Guinea.
Four intergovernmental agreements were signed on diplomatic relations,
cultural relations, shipping, and air transport. However, suspicion that
Cameroon was still less than totally committed to the Macias Nguema
regime persisted. Certain dissidents continued to urge annexation by
Cameroon; in 1977 Ekong Andeme, a former minister of health, and
other refugees called for annexation by Yaoundé.

As his government progressed, Macias Nguema increasingly sought
out contacts with the Eastern bloc and radical Third World regimes.
When, in 1971, the Organization of African Unity urged the extension
of aid to the republic, Libya responded by contributing $1 million to
Equatorial Guinea for the purpose of opposing "imperialist interference
and Zionist penetration of Africa."[20] Macias Nguema maintained particularly strong relations with the Congo People's Republic and visited
in August 1971.

Macias Nguema's overtures to the East and to radical African
regimes were intended to offset the threat of a coup or economic
strangulation. In 1970 Equatorial Guinea established links with the Soviet
Union. In June a trade agreement was signed that listed items on which
the two countries would apply most favored nation treatment. However,
the agreement was hamstrung by the provision that trade be conducted
in convertible currencies. The Soviet Union did obtain a base for deep-
sea fishing in Malabo and a naval base at Luba. The citizenry of
Equatorial Guinea was troubled because the Soviets were given a fishing
monopoly and fishing by nationals was forbidden. The People's Republic
of China also established relations in 1970 and subsequently extended
fifty-year interest-free loans. The Chinese aided in the development of
cotton and rice cultivation and supplied assistants for health and road
maintenance projects. In 1977 Nguema visited Peking and took the
occasion to criticize the Soviet Union. A bit closer to home, relations
with Cuba were good in the early 1970s but soured by 1977.

By the late 1970s, the president-for-life's rule was increasingly
distasteful to friend and foe alike, although neither took action to curb
the abuses. Terror within Equatorial Guinea demoralized internal op-

position; the country's isolation was acute. This was particularly true of Bioko, where the confiscation of canoes impeded escape to the mainland. A 1976 invasion attempt at Evinayong in Rio Muni was beaten back and resulted in reprisals. Around the same time, a group of government officials addressed a petition to the president-for-life, which asked for an amelioration of political and economic conditions. Many of the petitioners were executed, leaving only a clique of Mongomo Esangui as the president's intimates.

The Federation Internationale des Droits de l'Homme, the Anti-slavery Society, Amnesty International, the International University Exchange Fund, and the International Commission of Jurists finally took steps to expose the nature of the regime. In February 1978, the secretary-general of the United Nations attempted to sound out the government on the issue of human rights. When he received no reply, a working group of the Economic and Social Council (ECOSOC) recommended an examination of the situation. Subsequently, the Pan-African Youth Movement, meeting in Algiers, condemned the Malabo government for its violations of civil and human rights.

In March 1979 the U.N. Commission on Human Rights decided to name a special rapporteur on the situation. It further decided that the previously closed discussions would now become public. Also, in early 1979, a member of the World Labor Conference directing committee asked French President Valery Giscard d'Estaing to break off economic collaboration. Moreover, on his visit to China, Macias Nguema had been told that Chinese assistance would be cut back. Even friendly regimes appeared to tire of the mercurial behavior of their ally.

THE REGIME OF OBIANG NGUEMA

Macias Nguema was finally overthrown in a coup organized by his own subalterns. In late May 1979, public servants demonstrated in Malabo because of delayed paychecks. On June 4, five members of the National Guard demanded their wages and were shot. One of the executed military officers was the brother of Lieutenant Colonel Teodoro Obiang Nguema Mbasogo. The lieutenant colonel and a group of associates staged a successful coup on August 3. The leaders of the coup called on the Organization of African Unity, the United Nations, and Spain for assistance. Three days later an official Spanish mission arrived. In Spain, exiled members of Macias Nguema's regime (Masie Ntutumu, former minister of the interior; Ekong Andeme, former minister of health and later minister of industry and mining; and C. Mbomio, former chief of police) held a press conference and stated their support for the "anti-Communist" coup in progress.

The coup was somewhat protracted. Macias Nguema remained on the mainland when the revolt erupted. Surrounded by Guineanos trained in Cuba, North Korea, Romania, and China, he struck back militarily. He counterattacked near Monte Bata and then moved to Niefang, Nkue, Ebebiyin, and finally to Mongomo. Once he had been completely routed, the dictator fled into the forest in what appears to have been an effort to cross into Gabon.

Meanwhile, an initial 30 metric tons of Spanish material aid arrived and popular jubilation broke out in Malabo. The European Economic Community representative to Cameroon arrived and offered $10 million worth of assistance. The new head of state went to Libreville with the governor of Cameroon's Central Bank. Both Cameroon and Gabon seemed much more favorable to the new regime than they had been to its predecessor.

On August 18, Macias Nguema was arrested and taken to prison in Bata. There was some confusion about what to do with him. Obiang Nguema may have wished to have his uncle declared insane and exiled. Another associate in the coup, Maye Ela, Nguema's nephew and head of the navy, hoped for civil proceedings. From September 24 to September 28, a military court was conducted in a movie theater in Malabo. The first president of Equatorial Guinea was executed on October 1 after being tried for treason and genocide, among other crimes. In spite of his record, Macias Nguema was charged with only five hundred political murders. Only six other persons were tried and executed with him, including former Vice-President Miguel Eyegue.

Nine years after the death of Macias Nguema, one writer concluded that a "post-mortem of dictatorship in Equatorial Guinea is not possible."[21] The statement, however, may be somewhat too modest. It is possible to see the lines of continuity and discontinuity between the old and new regimes. A central issue has been each regime's need to justify its existence against the claims of more powerful neighbors and to appeal to powerful industrialized nations for financial and technical support. The foreign policy orientation of the government has shifted from left to right, but the state has not altered its extreme dependency.

Lieutenant Colonel Teodoro Obiang Nguema Mbasogo, the new chief of state, was thirty-seven years old and had received military training in Saragossa, Spain. Perhaps most importantly, he was not only a nephew of the executed Macias Nguema but had been vice-minister of defense under the old regime. Many doubted that his elevation to the presidency would change internal conditions. In December 1976 Obiang Nguema had been instrumental in the liquidation of the senior officials who petitioned the government about its disastrous economic policies.

Setting Up the New Regime

The new government promised both change and continuity. Its ideology was vague and its connections to the previous government obvious. The Mongomo "clan" retained its prominence and acted at points as a brake on the head of state. Relatives of Macias Nguema and functionaries in the old government remained in place. In addition to Obiang Nguema, Maye Ela, head of the navy and commander of Bata, and Ela Nseng, military governor of Rio Muni, profited from the demise of their kinsman and former leader. The government attempted to place all of the blame for the excesses of the previous regime on the Juventud. At the same time, much of the Juventud personnel was integrated into the post-1979 army. For example, Moro Mba, one of the main leaders of the Juventud, was made military commandant of Bata.

Obiang Nguema took power as head of the Supreme Military Council. With the passage of time, his government has taken on a more civilian cast. Former members of the armed forces who remain in the cabinet have ostensibly given up their military posts. In 1981 the president named a number of civilian commissioners to the council. A year later the government drafted a new constitution with the aid of representatives of the U.N. Commission on Human Rights. After a popular vote, the new document went into effect and the Supreme Military Council was abolished.

Under the new constitution, Obiang Nguema was given a seven-year presidential term. The powers of his presidency are considerable. He can make laws by decree, dissolve the legislature, negotiate and ratify treaties, call for parliamentary elections, and dismiss members of the cabinet. The office of prime minister was created and given the responsibility of coordinating matters of internal policy. The prime minister is appointed, dismissed, and granted power by the head of state. The president, the prime minister, the minister of defense, the chairperson of the National Social and Economic Council, and the president of the National Assembly are all members of a state council. This council is set up to act in place of the head of state in case of his death or incapacity.

By the end of 1983 the legislature was in place. The new body has 15 members appointed by the head of state and 45 chosen indirectly by the citizenry. All adult Equatorial Guineans have the right to vote for officials in their villages. The persons so elected in turn serve as electors for the national legislature. One person per district is chosen to serve in Malabo. The president appoints the seven provincial governors. The judiciary is headed by the president and his jurisconsults, who constitute the Supreme Court. Beneath this court are the appeals courts, chief magistrates for the districts, and local magistrates. Traditional laws are honored within the national court system provided that they do not conflict with state statutes.

The legislature, the Cámara de Representantes del Pueblo (Chamber of the People's Representatives), operates under the purview of the presidency and is not able to act without the president's sanction. Furthermore, legislators themselves have felt the weight of presidential displeasure. In June 1985 two legislators, Eduardo Ebang Masie and Antonio Ebang Mbele Abang, were charged before a military court with insulting the head of state. Ebang Masie was sentenced to eight and a half years in prison and fined for telling the people of his district, Añisok, that they did not have to applaud each time the president's name was mentioned. He also complained of interference in the activities of members of the Chamber. In October 1985 both he and Ebang Mbele Abang had their sentences suspended.

After eight years of rule Obiang Nguema and those around him formed a political party, the Democratic Party of Equatorial Guinea (PDGE). However, neither the constitution nor the legislature has defined the parameters of its political action. Obiang Nguema has promised that the new government party is only the continuation of a process of democratization. His political philosophy claims as its basis traditional African democracy: "National democratization based on the villages materializes the idea of the national political movement that should not be confused with the uniparty idea."[22] The people are, according to the president, at a stage preparatory to democracy.

The army and the police have been reorganized. The former metropole has trained hundreds of soldiers of all ranks. In late 1981 the Spanish inspector-general of police visited Malabo and agreed that Spanish soldiers would return to Equatorial Guinea as part of a large aid agreement. Arrangements were made to train two Equatorial Guinean companies, to be led by Spanish officers, in Spain.

Arbitrary acts by police and the military have declined. At the same time, the use of several hundred Moroccan troops as a presidential bodyguard has provided the leadership with a further hedge against a coup d'état or popular revolution. Although local forces outnumber these troops, they are ill-equipped. Of 1,395 military personnel in 1987, there were 1,100 in the army, 200 in the navy, and 95 in a small air force.

Government Abuses of Power

Government leaders have been able to extract economic benefit from both local and international capital. The head of state and his followers have used their political power to secure import licenses and

President Obiang Nguema, with visiting U.S. naval mission. (Photo courtesy Frank Ruddy)

other economic benefits. In April 1981 the government quashed a coup reportedly backed by Moses Mba Ada, businessman and former leader of MUNGE. In 1980 Mba Ada had formed a trading company called Exigencia, with José Rovira, a Spanish entrepreneur, and Justino Mba Msue, a local businessman. Mba Ada alleged from exile that the supposed putsch was an attempt by Obiang Nguema to gain complete control over the company. The head of state originally held 20 percent of the company, a share he later expanded to 50 percent through family intermediaries. As a result of the coup, Obiang Nguema received complete control of the firm. He also profited from the World Bank Cocoa Rehabilitation Project. In the early 1980s the bank announced that agricultural loans would be granted only to landowners. The president and members of his family were quick to acquire land. At present Obiang Nguema and his brother each own approximately 1,000 hectares of some of the best cocoa land on Bioko; the grants amount to 10 percent of all land redistributed by the regime.

The government did eliminate the most glaring abuses of the Macias Nguema period. The appointment of a Bubi vice-president, S. Seriche Bioco, signaled a desire to give the regime the appearance of a broader ethnic base. However, Seriche Bioco remained the only non-Fang in high office. Armengol Ondo Nguema, the president's elder brother, has been director of national security. The two other key security posts are in the hands of men with ethnic and personal ties to Obiang Nguema. Jesus Ngomo Nvono is head of the police and Ebengeng Nsomo is the deputy minister of defense. In a cabinet realignment in early 1986, Obiang Nguema assumed duties as minister of defense.

The new constitutional charter is criticized by the International Commission of Jurists as entrenching the powers of the president and facilitating the continued domination of the Mongomo clique.[23] Doubtlessly, the new government does represent a relaxation of the unremittent terror of the Macias Nguema years. However, the relaxation has only been relative. Forced labor continued early in the regime. Moreover, various infringements of political and social rights still exist. In 1983 the International Commission of Jurists informed the U.N. Commission on Human Rights that the country could only develop if greater popular participation was encouraged. After the promulgation of the new constitution, the commission made it very clear that much still remained to be done. The lack of the possibility of organizing an opposition was noted.

Furthermore, the commission of jurists noted that the construction of the constitution "reinforced the affirmations of the opposition that the true objectives of the present Government and of Colonel Obiang Nguema are to perpetuate themselves in power and to institutionalize a system that permits them to maintain total control over the political life of the country." The point was made "that, in spite of the fact that the population of Equatorial Guinea is formed from six different ethnic groups, the present President has entrusted the principal positions of the State administration largely to people of his town and home region (Mongomo)." There was concern that "this may create difficulties, affecting equality before the law and the prohibition of discrimination that the Constitution proclaims."[24]

Police brutality persists; the government arrested over 100 people after a purported coup attempt in July 1986. In the aftermath, 15 people were tried under the jurisdiction of a military tribunal. In October 1987, 6 of the supposed plotters were released as part of a government amnesty. Cases of brutality continue to be reported and, at its best, the regime can be characterized as one of *"ni sang, ni liberté."*[25]

The 1982 constitution provides for the right of association. However, this right, which would authorize political parties and unions, has not really been put into practice. In 1985, the U.N. Commission on Human Rights expressed concern that "human rights and fundamental liberties be fully and rigorously respected." In 1985 the ECOSOC designated an expert to "make sure that human rights and fundamental liberties would be fully and rigorously respected." In late 1985 a U.N. expert, Fernando Volio-Jimenez, and Obiang Nguema met in New York. They agreed on a visit by investigative jurists from Costa Rica between middle December 1985 and middle January 1986. The United Nations also decided on more specialists "in order to help the government . . . to elaborate other means to implement the constitution concerning the *effective* protection of fundamental human rights."[26] In March 1986 the commission on human rights asked the government of Equatorial Guinea to make the necessary amendments to its fundamental law: (1) to facilitate the repatriation of all the refugees and exiles, and (2) to adopt measures permitting the full participation of all Guinean citizens in political, economic, social, and cultural affairs.

The government and its supporters have been highly adept at obscuring the issue. The 1985 report of Amnesty International, published in early October, did not include Equatorial Guinea. It did say that the absence of certain countries was not proof of an absence of human rights violations but only indicated the difficulty of getting information. However, on October 30, *Jeune Afrique* wrote that "only three countries have the honor of not being on the list—Ivory Coast, Equatorial Guinea, and the Comoros." In November, *Paris Match* proclaimed that "the way is free for French enterprises. . . . Today under the guidance of Teodaro [*sic*] Obiang, faithful friend of France, democracy is reestablished." At the same time in Malabo, Luis Yanez, the Spanish secretary of state for international cooperation, emphasized that Equatorial Guinea was one of the African countries considered most respectful of human rights. In January 1986 the journalist Ndongo Bidyogo wrote from Malabo: "The success of the various freedoms was underlined in the last [Amnesty] report in which, according to *Jeune Afrique*, Equatorial Guinea is, with the Ivory Coast and the Comoros, among the only African countries that respect human rights."[27]

In May 1987, Amnesty International published a highly censorious report on the use of the death penalty in Equatorial Guinea. It urged the government "to ensure that all political prisoners receive a fair and public trial by an independent and impartial tribunal with full guarantees for their defence." In pursuance of this, Amnesty asked for the institution of "such changes in law and in practice as might bring procedures and practices in Equatorial Guinea into accordance with internationally recognized standards of fair trial." In particular, it recommended that torture and ill-treatment should not be used to force defendants to testify against themselves.[28] In its 1988 report the organization noted that Equatorial Guinea did not respond in detail to the complaints of the previous year. On the other hand, the republic did accede to both the International Covenant on Civil and Political Rights and its Optional Protocol. The latter permits the U.N. Human Rights Commission to receive complaints from individuals.[29] The covenant also contains guidelines for fair trials. At this point, whether Equatorial Guinea will fully adhere to this document remains to be seen.

Opposition to the New Regime

Both internal and external dissent continue to trouble Equatorial Guinea. One of its first internal manifestations occurred in February 1981. In Bata a student strike to protest continuing economic hardship resulted in the arrest and execution of adolescents. In 1983 Buale Borico, a high-ranking functionary and graduate of a Swiss technical school, fled to Spain after protesting continuing violations of human rights. In the same year there were reportedly two coup attempts. In the first, which took place in May, the alleged chief conspirator was Lieutenant Pablo Obama Eyang. Also implicated were Norberto Ela, the director of public works, Carmelo Owono Ndongo, a former commissioner to the presidency, and Jaime Obama Owono, the director of munitions. Suspicion also fell on Florencio Ela Maye, the ambassador to the United Nations.

The purported coup may have reflected the sad state of the economy. Carmelo Owono Ndongo and another accused plotter, Gregorio Micha, were sentenced and executed in Malabo. A third man, Sergeant Venancio Miko, fled to refuge in the Spanish embassy. President Obiang Nguema threatened to invade the embassy and only the appearance of Spain's foreign minister, Fernando Moran, averted a serious clash. Miko was turned over to the president's Moroccan guard and sentenced to death. On a later trip to Spain, Obiang Nguema promised that Miko would not be executed. However, according to exile sources, Miko had been executed with the knowledge of the Spanish embassy, which wished to maintain good relations with Obiang Nguema.

Late in 1983 there was a second purported coup attempt. Apolinar Boiche, the secretary of state for foreign affairs, was arrested along with other members of his staff. In 1985 Felix Mba Nchama, a former ambassador to Ethiopia and minister of interior, defected to Cameroon; the regime alleged that he had been involved in the 1983 plot against the chief of state.

Furthermore, in August 1986 Eugenio Abeso Mondu, a former military officer and diplomat serving in the Cámara de Representantes del Pueblo, was accused of attempting to overthrow the government. He was tried and publicly confessed to recruiting army officers for the assassination of Obiang Nguema. On August 19 he was executed by a military tribunal. Also tried was Fructuoso Mba Oñana Nchama, the deputy prime minister. Oñana Nchama was minister of defense until early 1986; at this time he was moved to the Ministry of Public Works following reports of unrest within segments of the army. He was sentenced to twenty-eight months in prison. Damian Ondo Mañe, national director of the Banque des Etats d'Afrique Centrale (BEAC), received the same sentence.

Although he has called for national reconciliation, Obiang Nguema has ruled out the participation of many opposition politicians. In November 1979 a special observer for the U.N. Commission on Human Rights said that returning refugees were being imprisoned, particularly in Evinayong. Exile leaders have continued to condemn the regime, although they have difficulty in speaking with a common voice. In 1983 there were approximately 25,000 Guineanos living abroad, many of whom had adopted a wait-and-see attitude toward the new government.

As in the period before 1968, the formation of political parties has involved personalities and ethnicity far more than strong ideological commitments. In 1980 a new opposition group, the Democratic Movement for the Liberation of Equatorial Guinea, was launched by Manuel Ruben Ndongo in Calabar and appealed to the Organization of African Unity for aid. In August 1981, representatives of two wings of the Alianza Nacional de Restauración Democrática (ANRD) and those of a more radical group met clandestinely in Rio Muni and selected former minister Daniel Oyono to head a new opposition front. In September Oyono announced the creation of a new opposition in Paris, an action followed by the arrest of twenty people in Equatorial Guinea. Another exile group, the Progress Party of Equatorial Guinea (Partido de Progreso de Guinea Ecuatorial, or PPGE), is headed by a former minister of information, Severo Moto Nsa. Yet another group, the Convergencia Social Democrática (CSD) is, like the PPGE, basically a Fang grouping with an audience among exiled Guineanos. The Frente de Liberación de Fernando Póo is Bubi-based and advocates separate independence for Bioko.

In April 1983 several of the groups came together in the Junta Coordinadora de las Fuerzas de Oposición de Guinea Equatorial (JCFOGE). All of these groups, whether based in Spain, Switzerland, or elsewhere, are underfunded and rent by factionalism. Obiang Nguema uses a carrot-and-stick approach toward the opposition, that thus far has paid dividends. He has welcomed back some exiles like Samuel Ebuka, early postindependence ambassador to Nigeria. At the same time he has shown himself very much opposed to power-sharing with dissidents. In the summer of 1983 the president of Equatorial Guinea visited Spain but dashed hopes for a reconciliation with political exiles. Obiang Nguema called reports of political murders and the mass exodus of citizenry "mere rumour" and labeled the opposition as "illegitimate." Although the government put forward vague proposals for "political reforms," he did not give any indication as to its timetable. A 1985 offer of negotiations from Obiang Nguema was subsequently rejected by two exile groups who felt it did not go far enough.

In June 1988 Moto Nsa of the PPGE returned to prepare to contest the 1989 presidential elections. He was accompanied by the party's secretary-general, José Luis Jones, and André Louis, the assistant secretary-general of the Christian Democratic International. Moto Nsa had taken the precaution of soliciting the support of the three main Spanish opposition groups: Alianza Popular, Centro Democrático y Social, and the Partido Democrata Popular. They established an oversight committee to monitor future elections. A delegation from the European People's party, the Christian Democratic party in the European parliament, also went to Malabo later in the summer. It offered its own carrot and stick. If democratization took place, it offered the promise of a lobbying effort in Europe on behalf of aid to Equatorial Guinea.

Initially, Obiang Nguema appeared to agree with the idea of contested elections. On the delegation's return to Europe, he let it be known that he would not recognize any opposition party. He did note that the PPGE could be recognized under Equatoguinean law if the members of the legislature approved. As members of the only extant party, the president's PDGE, it is very doubtful that they would approve. In spite of promises of a return to multiparty rule, in September 1988, Obiang Nguema arrested returned members of the opposition, a move that seemed to portend a one-party election. The government announced that it had discovered yet another coup. As a result, a wave of arrest occurred; Spain made a formal protest after three Guineanos with Spanish passports were imprisoned and held incommunicado in Bata. Later, two members of the military were executed and the government blamed the PPGE for causing discontent. In 1989, to the surprise of few, Obiang Nguema was chosen for a second seven-year term.

In the autumn of 1988 José Luis Jones, the secretary general of the PPGE, was released from prison after being condemned to seventeen years in prison. To some members of the exiled opposition, this was an example of too little, too late. In March 1989 Equatorial Guinean émigré groups and representatives of most Spanish parties signed a protest demanding open elections and the freedom of all political prisoners. Support for the statement also came from the Catholic church. Of the political parties, only the socialists of the Premier Felipe González demurred. This may have been an attempt to retain leverage with Malabo, in spite of a rising chorus of Spanish complaints about Equatorial Guinean human rights violations.

Relations with Other Countries

Western opinion does have an impact in Malabo. The head of state is as uncritically pro-Western as his predecessor was pro-Eastern bloc. Although the Soviet Union lost some political leverage with the demise of the old regime, it continues to maintain high visibility in Malabo and has sold the republic oil at lower than world prices. Relations continue to be cordial; in August 1988 a parliamentary team from Equatorial Guinea was invited to Moscow as guests of the Supreme Soviet. In addition, the People's Republic of China has fielded a technical aid mission of several hundred and reopened the central telephone exchange. The entrance of Equatorial Guinea into the Communauté Financière Africain (CFA) zone in the mid-1980s has increased links with Cameroon. Managerial personnel and goods are entering the country through Douala. If Equatorial Guinea becomes an important food exporter, the obvious ties between the two economies will have political ramifications.

For more than forty years, Equatorial Guinea's most important relationship has been with its northern neighbor, Nigeria. Under colonialism this relationship was the key to the success of the export economy. At present the World Bank has a program to rehabilitate it. By the mid-1980s cocoa production had again fallen below the levels sustained in the middle years of Macias Nguema's rule. Obiang Nguema visited Nigeria in 1981 and 1982 in an effort to resume labor migration. In 1985 further attempts to secure any labor agreement fell through. Cases of labor abuse in the same year did little to enhance Equatorial Guinea's image as a desirable place of employment.

After 1979 Lagos watched the return of the Spanish presence with suspicion. Spain's potential membership in NATO was not looked on with favor, especially as NATO interests may run counter to Nigerian interests on such issues as southern Africa. In 1981 Shehu Shagari emphatically emphasized the Nigerian navy's role in protecting neighboring countries against subversion, a message most applicable to the insular portion of Equatorial Guinea. Even before the advent of the present regime, an adviser to the Nigerian government observed: "It is not going to be difficult for Nigeria to deal with either China, the Soviet Union or the United States, firmly over Fernando Po [Bioko]. There might come a time when America's dependence on oil exports from Nigeria might be used as quid pro quo to their withdrawing from Fernando Po." Furthermore, "the Chinese and the Soviets are quite aware of the potential influence of Nigeria in Africa and they are not going to forfeit their friendship for a transient state like Equatorial Guinea."[30]

Concern was doubly heightened in the late 1980s when South Africa showed an interest. As early as 1985 reports began to circulate that Pretoria had a seven-person team in Equatorial Guinea. It appears that South Africans were at work extending the runway at the Malabo airport. They also introduced 150 head of cattle at Moka on Bioko. The cattle ranch was run by Oprocage, a company headed by Hilton Lack, a senior civil servant in the South African Ministry of Foreign Affairs. The white minority regime also supplied Equatorial Guinea with cheap copper sulphate, an important fungicide. South African C-130 transports made frequent night visits to Bioko and reports circulated that a satellite-tracking station was being built. The South African government also reportedly granted $20 million to Equatorial Guinea. After the visit of a group of South African businessmen in 1987, an American journalist concluded that "their [the South Africans'] presence in this former Spanish colony is part of a new South African policy to win friends on small African islands with big airfields. . . . On the East coast of Africa, the South Africans have started similar cattle breeding projects in the Comoros Islands and on Mauritius."[31] Such activities, combined with South African strategic interest, especially concerned the Nigerian government.

Early in 1987 Obiang Nguema visited Lagos and denied that South African nationals were operating in his country. He said that "by no means will my country have a pact with the racist country and we have manifested this in all our actions."[32] He also rejected a nonaggression pact with his host. In addition to helping build up the Equatorial Guinean navy, Lagos approved a loan of 5 million naira at 3 percent interest to be repaid in ten years. Nigeria offered its neighbor the services of its Technical Aid Corps and Nigerian businessmen were encouraged to invest in Equatorial Guinea's agricultural sector. The entrepreneurs agreed to provide backing for the manufacture of plastics. Discussions were also undertaken on the rehabilitation of Malabo's hotels and the construction of medical facilities, recreational centers, and schools.

Nigerian suspicion of Pretoria's presence in the Bight of Biafra continues. There is fear that several powers, including Cameroon, may be fishing in troubled waters. President Paul Biya is not as Francophile as his predecessor; this is especially true after an abortive coup attempt by ex-president Ahmadou Ahidhjo, who was accused of having French backing. One Nigerian expert suggests that Cameroon may wish to replace France with Israel as its major external prop.[33] A Cameroon-Israeli military agreement was signed following the visit of Shimon Peres to Yaoundé in 1986. According to this scenario, the South African presence in Equatorial Guinea may have something to do with the joint economic and strategic aims of Pretoria and Jerusalem. This argument rests on a congruence of interests that has yet to be proven. By the same token, West German interests have also been said to be involved. A German electronics firm, Siemens, was to have produced telecommunications equipment for the satellite-tracking station on Bioko. This information was contained in a leaked report prepared by the Hans Seidel Stiftung, a conservative research institute aligned with the late Bavarian politician, Franz Josef Strauss, described as an "open ally of Pretoria."[34]

In addition to other activities, Equatorial Guinea has supposedly resold Nigerian oil to South Africa. In May 1988, General Ike Nwachukwu, Nigeria's external affairs minister, visited Malabo and strenuously expressed his government's concern: "We told Equatorial Guinea that South Africa's presence is unacceptable to Nigeria since it is a threat to our security." Obiang Nguema admitted that South Africans were in his state but said that they were only there in a private capacity. Shortly before a mid-May meeting of the OAU's Council of Ministers in Addis Ababa, Equatorial Guinea claimed to have expelled all South Africans. Obviously, the move was intended to ward off criticism from the OAU and Nigeria. Shortly before the organization's heads of state meeting in late May, President Ibrahim Babangida of Nigeria noted that "some fellow African states are being lured into contrived friendship or subtle collaboration with the Pretoria regime."[35] In January 1990, he visited Malabo in an effort to persuade Equatorial Guinea to adopt policies more in line with those of its African neighbors.

As in the past, Nigerian anger at its neighbor seemed to be mixed with much caution. The Nigerian foreign minister, Alhaji Mamman Anka, said that the reported expulsion of the South Africans had "amicably settled" the issue.[36] However, the South African presence and the issues it raises have not rested. In the summer of 1988 relations between Malabo and Lagos were tense because of continued sightings of South African personnel. Also, the South African government condemned Nigeria's attitude and said that it would continue its projects in Equatorial Guinea. In June 1988 the Obiang Nguema government protested to the United Nations that Nigeria was interfering in its internal affairs. The government of Equatorial Guinea also expelled a Nigerian diplomat. Some members of the Nigerian foreign policy apparatus urged a strong interventionist stand; others stood firm against any new acquisition of territory.

Future Politics and Interests

Even if Nigeria should mute its voice, Equatorial Guinea will still face other African criticism. In October 1988, *L'union*, a government-backed Gabonese newspaper, chided: "Confronted with a powerful and intransigent enemy, the dialogue looks like the weapon of the weak. One cannot talk Pieter Botha into a sort of self-denial, which is demanded by all, unless various pressures are brought to bear on him both at the continental and international levels."[37]

Obiang Nguema's government sees itself as a pauper taking help from whomever will give it. In 1988 this was amply demonstrated when the government agreed to receive at least 5 million metric tons of toxic industrial waste from Europe. The wastes are to be deposited on Annobón. Obiang Nguema approved of the project in return for a down payment of $1.6 million. It is projected that a British company will transfer 2 million drums a year to the tiny volcanic island. The president is also charged by the opposition with signing a ten-year agreement with the U.S.-based Axim Consortium Corporation to dispose of 720,000 tons of toxic waste per year.

To outsiders it appears that the government has substituted an opportunistic pro-Westernism for the erratic anti-Westernism of the pre-1979 period. In 1988 it was rumored that Malabo was moving closer to Israel, perhaps in a bluff to increase the contributions of various Arab states. Morocco said that it would withdraw its presidential guard if Obiang Nguema took such a step.

If Obiang Nguema cannot sustain the benefits flowing to his followers and colleagues, especially those in the Mongomo clique, he may well find himself cut off from a needed base of support. Tensions within the regime persist; according to some observers, the Mongomo clique has acted to check the Western orientation of the regime. In September 1981 the head of state publicly accused certain Mongomo personages of economic sabotage. In the ensuing power struggle, the Mongomo group reportedly emerged victorious. In 1988 Mba Oñana, the uncle of President Obiang Nguema who had been imprisoned for supposedly supporting the 1986 coup attempt, was released from prison. Mba Oñana is a member of a hardline faction within the Mongomo group and is popular with the army. His release could represent a reconciliation between Obiang Nguema and his kinsmen. At any rate, it is a recognition of their continuing importance.

Equatorial Guinea has been called a "dictature de type haitien."[38] The events of 1979 did not result in democracy. Indeed, the president has made it quite clear that his country is not ready for democracy because "there is ignorance among Africans as to what democracy is."[39] Obiang Nguema maintains an authoritarian regime that endures because of the sufferance of aid donors more concerned with possible strategic and economic benefits than with human rights. Although conditions have changed, Equatorial Guinea is no closer to democracy now than it was in the 1970s.

It has been said that because it is "an island and mainland divided by 150 miles of ocean, with two conflicting ethnic and linguistic groups and little sense of nationhood, Equatorial Guinea clearly needs strong leadership. This it has."[40] Still, authoritarian rule is not necessarily stable rule. Obiang Nguema will have extreme difficulty in the future if he cannot bring dissidents into the political arena while maintaining a system of rewards and benefits for his associates. It is within this group of associates that real destabilization of the regime may come.

NOTES

1. Brian Weinstein says it was Ndongo. Brian Weinstein, *Gabon: Nation-Building on the Ogooué* (Cambridge, Mass.: MIT Press, 1966), p. 228.
2. René Pélissier, "Political Movements in Spanish Guinea," *Africa Report* (May 1964), p. 6.
3. René Pélissier, "Uncertainty in Spanish Guinea," *Africa Report* 13, 3 (March 1968), p. 19.
4. *Ibid.*
5. Robert af Klinteberg, *Equatorial Guinea—Macías Country* (Geneva: International University Exchange Fund Field Study, 1977), p. 43. Klinteberg says that Macias was born in the village of Oyem in the Woleu-Ntem province of northern Gabon. It is also said that he was born in the village of Nsangayong in the Mongomo district. Klinteberg, a Swedish anthropologist, made a fact-finding tour of Equatorial Guinea and its neighbors (Gabon, Cameroon, and Nigeria) in 1978. His tour was sponsored by the International University Exchange Fund, with aid from the Swedish International Development Authority (SIDA).
6. Robert Gard, "The Colonization and Decolonization of Equatorial Guinea" (Pasadena, Calif., unpublished manuscript, 1974), p. 330, citing "Conferencia Constitucional," Document No. 2: Informe sobre Guinea Ecuatorial de los Miembros de la Comisión Interministerial, vice-presidencia del gobierno.
7. Billy Gene Hahs, "Spain and the Scramble for Africa, the 'Africanistas' and the Gulf of Guinea" (Ph.D. diss., University of New Mexico, 1980), p. 2.
8. Buenaventura Ochaga Ngomo, "Nacimiento de la libertad de Guinea Ecuatorial," *Organo informativo del Ministerio de Educación Nacional de Guinea Ecuatorial* 7 (March 8, 1970), n.p.
9. Alejandro Artucio, *The Trial of Macias in Equatorial Guinea: The Story of a Dictatorship* (Geneva: International Commission of Jurists and International University Exchange Fund, 1980), p. 2.
10. Alianza Nacional de Restauración Democrática, Position Paper (Geneva, 1979), p. 8.
11. Max Liniger-Goumaz, "La République de Guinée Equatoriale," *Acta Geographica*, 3d ser., no. 43, Société de Géographie, Paris (1980), pp. 1–22.
12. *Ibid.*, p. 46.
13. Artucio, *The Trial*, p. 10.
14. Klinteberg, *Equatorial Guinea*, p. 33.

15. *Ibid.*, p. 84, app. 1, extract from a speech by Macias at the constitutional conference, November 3, 1967.

16. Bolaji Akinyemi, "Nigeria and Fernando Poo, 1958–1966: The Politics of Irredentism," *African Affairs* 69, 276 (July 1970), p. 244, citing *House of Representatives Debates*, April 3, 1962, p. 30.

17. "Update," *Africa Report* 12, 1 (January 1967), p. 23.

18. *Africa Report* 14, 2 (March–April 1969), p. 31, citing *Le Monde* (Paris), January 22, 1969.

19. *African Contemporary Record, 1972*, ed. Colin Legum (London: Rex Collings, 1973), p. B548, henceforth cited as *African Contemporary Record.*

20. *African Contemporary Record, 1971*, p. B505.

21. Samuel Decalo, *Psychoses of Power* (Boulder, Colo.: Westview Press, 1988), p. 69.

22. Teodoro Obiang Nguema, *Guinea Ecuatorial, País Joven* (Malabo: Ediciones Guinea, 1985), p. 103.

23. Naciones Unidas, Consejo Económico y Social, Comisión de derechos humanos, 39° periodo de sesiones, tema 12 del programa, *Comunicación escrita presentada por la Comisión International de Juristas, organización no gubernamental reconocida como entidad consultiva de la categoría II. Nueva Constitución de Guinea Ecuatorial* (E/CN.4/1983/NGO/4), January 31, 1983.

24. Marcos-Manuel Ndongo, "Guinea Equatorial, Posibilidades de desarrollo económico," *Africa 2000*, año 3, epoca 2, num. 6, p. 23.

25. Welter Hink, "Ni sang, ni liberté," *Afrique-Asie* 238 (April 27–May 10, 1981), n.p.

26. Max Liniger-Goumaz, *Connaître la Guinée Equatoriale* (Geneva: Editiones des Peuples Noirs, 1986), p. 204.

27. *Ibid.*, p. 203.

28. Amnesty International, *Equatorial Guinea: Military Trials and the Use of the Death Penalty*, AI Index, no. AFR 24/01/87 (New York: Amnesty International), p. 12.

29. Amnesty International, *Amnesty International Report, 1987* (London: Amnesty International Publications, 1988), p. 38.

30. Akinjide Osuntokun, "Nigeria–Fernando Po Relations from the Nineteenth Century to Present" (Paper delivered at the Canadian African Studies Association Conference, Université de Sherbrooke, Quebec, April 26–May 3, 1977), p. 171.

31. James Brooke, "Pretoria Lends Helping Hand to Friends with Big Airstrips," *New York Times*, October 21, 1987.

32. "Dagger at the Underbelly," *Newswatch* (Nigeria), May 23, 1988, p. 15.

33. Akinjide Osuntokun, "The Security Aspect of Nigeria–Equatorial Guinea Relations" (Paper presented at the Nigerian Institute of International Affairs, Lago, April 28, 1987), p. 15.

34. Ernest Harsch, "Botha Secretly Attempts to Forge Alliances with Independent African States," *The City Sun*, June 14–21, 1988, p. 8.

35. "Update," *Africa Report* 33, 4 (July–August 1988), p. 11.

36. Earnest Harsch, "Country Briefings," *African Business* (August 1988), p. 36.

37. Foreign Broadcast Information Service, *Africa Daily* (Daily Report, National Technical Information Service, Springfield, Va., November 2, 1988), p. 3.

38. Max Liniger-Goumaz, "Deux ans de dictature post-Macias Nguema," *Genève-Afrique* (Journal of the Swiss Society of African Studies) (1981), p. 177.

39. *Courier* (Nigeria) 107 (January–February 1988), p. 40.

40. *African Contemporary Record, 1985–1986,* ed. Colin Legum (New York: Africana Publishing Company, a division of Holmes and Meier, 1987), p. B222.

4
The Economy

In 1968 Equatorial Guinea was put forth as a model of colonial development. Soon after, the highly deceptive colonial economy collapsed. It is estimated that cocoa production fell from 38,300 metric tons at independence to approximately 6,670 metric tons in 1978–1979. Except for coffee (a few hundred metric tons), the export economy had collapsed by the early 1979.[1] Subsoil wealth remained unexploited and the agricultural sector was disrupted. Terror interfered with the migratory labor system upon which the plantation economy depended. Without this system, any idea of continued economic viability based on export production, much less an independent monetary system, became illusory. By the time the regime collapsed, Equatorial Guinea had a largely barter economy. Its currency was nonconvertible and no rules existed for investment, economic policies, or commercial practices. The decline of the economy and erratic brutality combined to undermine the developmental schemes characteristic of the first years of independence.

One consequence of the brutality of Macias Nguema's regime is the lack of economic data. As members of the educated population, statisticians were subject to liquidation. By the late 1970s almost nothing was known about the economy in quantitative terms. The U.N. Development Program considers statistics-gathering to be one of its priorities. Rehabilitation in this respect is still incomplete. A 1988 statistical survey noted that "official statistics are rare, confusing and very late in publication."[2]

DECLINE

Given the political course of Equatorial Guinea during the Macias Nguema years, it is not surprising that both exports and imports declined between 1968 and 1970. Unfortunately, the former declined much faster than the latter. This trend was already obvious in the preindependence period; from 1966 to 1970 imports of mineral products, tobacco, and

91

clothing increased greatly. On the other hand, imports of transport vehicles and construction materials decreased. Spain remained the chief source of imports and the chief destination of exports. Six years before independence, the metropole took 98 percent of the colony's exports and provided 68 percent of its imports.[3] A year after decolonization, 91 percent of exports went to Spain and 80 percent of imports came from Spain. Most of the remaining exports went to Great Britain.

Spanish Economic Participation

As mentioned earlier, the decolonization envisioned by Madrid in the late 1960s included a healthy dose of Spanish economic participation. In May 1969, Spain and Equatorial Guinea signed economic agreements that promised to benefit both sides. Macias Nguema received the right to establish a bank of emission under extremely liberal terms; the conversion rate of the national currency was to be determined by the African state. The Spanish Institute of Foreign Money was required to open a dollar account for the National Bank of Equatorial Guinea. The institute was to keep track of the dollar value of Spanish exports to Guinea and of Guinean exports to Spain. The amount of indebtedness that either party could incur was set at $5 million. In the event that the balance of either account showed a deficit in excess of this amount, or in case the agreement was voided, the debtor nation would "proceed, at the solicitation of the other party, to pay the difference to the creditor country in any convertible currency determined by mutual agreement of the respective Institutions at that time."[4]

The agreement stipulated that Spain buy at least 20,000 metric tons of cocoa per annum, 1,850 metric tons of cocoa products, 6,000 metric tons of coffee, 215,500 metric tons of wood, and lesser quantities of other produce. The agreement also bound Equatorial Guinea not to export these agricultural products to other countries before the Spanish quotas had been met. Spain agreed to purchase Equatorial Guinea's produce at higher than world prices. Macias Nguema's government, in return, promised to give preference to Spanish goods, if they were competitive in price and quality with goods originating elsewhere. In 1970, the last year of the Macias Nguema regime for which detailed statistics are available, 80.46 percent by value of all imported goods were Spanish; 90.5 percent by value of Equatorial Guinea's total exports went to Spain, including 86 percent by value of the total amount of cocoa exported and all of the coffee and timber exported. Spanish–Equatorial Guinean cooperation was also promised in other areas. Spain consented to allot a total of 7 million pesetas in educational grants to Guineans and to provide 100 teachers to establish schools in Bata and Malabo.

The attempted coup of 1969 soured relations. The big Spanish companies were closed. European skills and capital fled the nation and crisis set in. In 1969, the secretary-general of the United Nations sent a personal representative, Marcial Tamayo of Bolivia, to survey the economic health of the new republic. At that time, it was apparent that there was an urgent need for financial assistance and replacements for Spanish civil servants, technicians, teachers, doctors, and merchants. In 1969 the World Health Organization (WHO) sent a team of advisers, along with other agencies of the United Nations. In the same year, the Organization of African Unity described the country as on "the brink of social, economic and political chaos," and "in the most inextricable disorder."[5] An appeal by the OAU produced some favorable responses, but they were unable to produce a basic shift in the economy.

Relations were difficult following the attempted coup, but a rapprochement seemed in the offing in 1971. Jesus Oyono, minister of public works, visited Madrid and stated that talks had been conducted in a spirit of "total sincerity, trust and friendship, and were motivated by the desire to start a new era in relations."[6] Both sides agreed on the construction of a training school for industrial specialists, economic and technical coordination, and the reopening of the TV transmitter in Malabo, which had been closed since the exodus of Spanish technicians in 1970. Further cooperation was promised in marine and air transport.

Economic agreements between Equatorial Guinea and Spain were renewed in 1971 and 1973. Until January 1972, Spain maintained highly preferential tariff arrangements with the former colony. In 1975 the former metropole was reportedly giving Equatorial Guinea considerable budgetary aid. This was in addition to "cultural" aid, which was 7 million pesetas ($108,000) per annum. In the same year, Madrid granted 50 million pesetas ($780,000) per annum in credits for a three-year period, chiefly for buying equipment for cocoa plantations.[7]

Ultimately, Spanish efforts to aid development failed. Spanish interests could not overcome Macias Nguema's almost innate suspicion. Furthermore, the Spanish were continually threatened by the president's attempts to establish a liaison with non-Spanish capital. In 1972 Spanish commerical banks withdrew from the country because of a deteriorating economic and political situation. By 1978, economic relations with Spain had almost come to a standstill. The continued presence of Spanish interests had been only at the sufferance of the president, who was increasingly unimpressed by Madrid's efforts to placate him.

Before independence, Macias Nguema had been the candidate favored by two sections of Iberian capital: construction companies and "new" Spanish capital that hoped to break the oligopolies dominant in the preindependence period. In the long run, the latter proved to be more important. Macias Nguema's rise was abetted by Antonio Garcia-Trevijano, a lawyer and financier. Reputedly connected with Opus Dei, he represented the technocratic and "modernizing" aspect of a section of Spanish capital. Before 1968 Garcia-Trevijano had made himself the Maecenas of several political factions before settling on Macias Nguema. The Spaniard urged that capital be procured from countries other than Francoist Spain and saw himself as the facilitator of this new influx. At the end of 1969 the Instituto de Fomento de Guinea Ecuatorial (INFOGE), an organization with a monopoly on foreign trade, was set up at his suggestion. INFOGE was placed under the directorship of Moses Mba Ada and was intended to take the country out of the Spanish economic sphere. It collapsed in 1972 and was replaced in 1975 by SIMED, SA. SIMED did not have a monopoly but had to face competition from members of Macias Nguema's family.

Even at the height of his terror, Macias Nguema was able to gain some backing for his rapidly unraveling economy. In 1975 the country joined the African Development Bank. When the European Economic Community refused to finance cocoa plantations because of human rights violations, the African Development Bank agreed to provide $8.9 million in 1978. Equatorial Guinea also received a loan of $500,000 from the Arab Aid Funds to Africa.

With the collapse of the banking system and the deterioration of record keeping, there were no instrumentalities capable of managing expeditures. In 1971 the treasury deficit amounted to $12.5 million. In the eyes of many, the priorities of the government were skewed. In the budget of 1972, foreign affairs, defense and interior, and the executive received 56 percent of government revenues. Technical ministries received 44 percent; health received 16.4 percent and education 10.5 percent. Government revenues were sustained by taxes on real estate and exported profits (20 to 35 percent). After 1970, public investment was channeled into prestige projects like the presidential palace at Ekuko near Bata and two deluxe buildings for the central bank and the port of Bata. Construction was one segment of the economy that prospered while the rest of the economy declined.

In terms of foreign capital, certain French interests responded to the diminution of the Spanish role with alacrity. France was the only Western country to maintain an embassy throughout the Macias Nguema years (e.g., the United States closed its embassy in 1976). France was especially interested in gaining control of Rio Muni timberlands. The opening of an embassy in Malabo, according to a French business journal, allowed "our private sector, with the support of export guarantees, to counterbalance the influence of Eastern countries, most notably the USSR, always on the search for fishing bases."[8]

The encouragement given to French forestry interests reflected the parlous state into which lumbering fell after independence. Timber production, which amounted to 100,000 metric tons in 1970, had come to a virtual halt by 1971. Some of the lands abandoned by Spanish companies were turned over to Equatorial Guinea's national bank. The machinery left behind by Spanish forestry concerns was placed at the disposal of the National Forestry Institute in Mbini (Rio Benito). In 1974 the U.N. Development Program reported that the forestry institute had "thus far been unable to take the necessary measures to organize the forestry industry, and today, after a lapse of five years, much of that equipment has deteriorated. The same is happening with the timber already cut and ready for export."[9]

In 1972 a Franco-Swiss firm, Societé Forestière du Rio Muni, was given title to the largest concession ever granted. It received a ten-year concession to some 150,000 hectares of virgin forest. Until 1974 the firm was able to export 400,000 metric tons of lumber annually. Because of the insecure political climate, the company sold out to another French concern, Tardiba, which almost immediately suspended operations. Another example of the penetration of French capital was the Societé Française des Dragages et de Travaux Publiques, which engaged in construction work in Bata and Malabo.

The Failure of Capital and Labor

Overall, infrastructure declined. New German generators were installed in Bata's thermal electric station in 1972–1973. Unfortunately, because of a lack of qualified personnel, the installation ceased to operate in 1976. A year later the Chinese contributed to the construction of a hydroelectric power station at Bikomo Falls, 15 kilometers from Bata. Meanwhile the thermal power station in Malabo had broken down. Personnel from the People's Republic of China constructed a 22.5-kilometer road from Nkue to Mongomo between 1972 and 1974. The U.N. Development Program, in collaboration with other bodies, started a $2 million project in 1973 to train road maintenance personnel. The project fell through; its director was ejected from the country in 1975.

Transportation also deteriorated. In 1971 a Hispano-Guinean agreement called for the creation of a national airline, Lineas Aereas de Guinea Ecuatorial (LAGE). The former colonial power contributed aircraft, pilots, technicians, and training scholarships for airline personnel. The Spanish government also provided a subsidy to its airline, Iberia, to provide service from Madrid to Malabo. It did the same thing via-à-vis sea transport between Europe and Equatorial Guinea. A subsidy was given to the Transmediterránea Company. Toward the end of the regime, the company's service became very sporadic. Telecommunications tell the same story of deterioration. In 1968 the government took over the communication installations of Torres Quevado, SA. In 1969–1970 Equatorial Guinea attempted to reestablish the communications network with the aid of Spain and the People's Republic of China. Help was also solicited from the United Nations. Later Cameroon financed a telecommunications link between Malabo and Yaoundé.

Macias Nguema hoped that U.S. and other capital would facilitate the exploitation of subsoil resources. Because of the need to find a new source of wealth, he allowed Western oil companies to explore in offshore waters. In 1969 the Continental Oil Company (Conoco) acquired onshore and offshore exploration rights. Conoco got the concession from Gulf Oil Company of Equatorial Guinea and from the Compañia Española de Petroleos de Guinea Ecuatorial, SA (CEPSA), each of which retained a 25 percent interest in the explorations. Continental was the operator for the group and promised to invest $2 million in exploration during 1970. This included the drilling of an exploratory well, in addition to geological and seismic studies. Shell Oil also maintained a Netherlands-based company known as Shell Guinea Equatorial.

In 1970 Chevron Petroleum did investigation and analysis in an area close to Ntem and Bata. The company gained a concession for 233 square kilometers. Income from the concession of prospecting rights provided a budgetary surplus of $1.2 million in 1969 and the following year contributed to a deficit reduction of $2.8 million. Unfortunately for the regime, these actions did not produce rapid results. The erratic behavior of the government did not inspire confidence and by the mid-1970s Equatorial Guinea still could not exploit the oil wealth that had proven such a boon to nearby Nigeria and Gabon.

Equatorial Guinea based its hopes for economic and political independence on the continued profitability of cocoa, its major crop. In 1972 it accounted for the majority of total exports. However, an exodus of both management and labor caused a drastic fall in production. On Bioko the postindependence mass exodus of many Spanish landowners threatened to collapse the economy. Plantation work was disrupted and many workers (reportedly 15,000 on Bioko) lost their jobs. In 1969–1970, the final cocoa production figures were 28,950 metric tons as compared to 38,300 metric tons for 1968–1969. In 1970–1971 the cocoa harvest showed some resiliency: 30,000 metric tons were harvested. In 1970–1971 Equatorial Guinea was Africa's fifth largest producer of cocoa. In 1972–1973 cocoa production was 13,300 metric tons. The figure had fallen to half of the 1970–1971 level of 30,000 metric tons and far short of the 1971–1972 estimate of 22,000 metric tons.[10] Coffee, the second major export crop, declined significantly after 1968. In 1972 production was 7,000 metric tons. Six years later it was one-tenth of that amount.

More important was the precarious nature of the labor supply. In 1970, Nigeria asked whether Equatorial Guinea would revise existing labor laws. The smaller country promised to remove all oppressive legislation and improve conditions. In the last month of the year, the Equatorial Guinean ambassador to Nigeria, Samuel Ebuka, met to discuss the matter with Chief Anthony Enahoro, the Nigerian commissioner for labor and information, but Nigeria remained wary of labor migration. On the contrary, the situation was exacerbated by the departure of 15,000 to 20,000 of a total of 40,000 Nigerian plantation workers.

The end of the Nigerian civil war in 1970 gave impetus to this movement. More importantly, the slowdowns in the whole cocoa economy resulted in the nonpayment and disaffection of workers. At the end of 1971 Macias Nguema's ambassador to Nigeria lamented that cocoa production had fallen off because hundreds of Nigerians had not recontracted for Bioko. He attributed this state of affairs to the dilatory way in which Nigeria contemplated ratifying a new labor agreement. This new agreement increased wages, provided free housing, medical attention, and food, and rigid regulation of working hours.

The postindependence series of labor treaties had fallen into disuse by the mid-1970s. Relations between workers and the government had become extremely tense in 1972 when approximately 50 Nigerian workers were killed after demonstrating over the arrest of fellow laborers. Following this incident, workers again began to desert the island. By the end of the year only about 12,000 workers were recruited for the cocoa plantations and there was friction between migrant and local workers. A group of Nigerians returning to Calabar claimed they had been forced to abandon their property. In the following year arrests and expulsions of Nigerians from Equatorial Guinea caused Lagos to suspend further recruitment. After the expiration of the bilateral labor agreement in February 1973, Anthony Enahoro visited Malabo and concluded that the terms of previous labor agreements had not been met. In July the Nigerian government refused to send 1,000 workers recruited in Calabar pending a full investigation and a new agreement. In 1974 Nigeria reviewed its labor dealings with Equatorial Guinea and noted that laborers had not been paid for periods longer than six months and that the local gendarmerie frequently attacked workers. In 1975 and early 1976 the Nigerian government airlifted thousands of its nationals out of Equatorial Guinea. The exodus gave the *coup de grace* to an already desperately ill economy.

As early as 1973 Macias Nguema's party proposed to remedy the labor crisis through the recruitment of 60,000 Equatorial Guineans. The following year the party noted that these workers had not been recruited and urged renewed efforts at production for export. Agriculture continued to run down. An African journalist reported that "in the countryside, cash crops have been replaced by fields of indian hemp and opium as the regime's answer to economic paralysis."[11] In 1976 the president-for-life decreed compulsory labor for all citizens over age fifteen. A year later there were 45,000 unpaid workers (including dependents), a figure representing about one-fifth of the population remaining in the country.[12] Labor protest was considered antistate activity.

Fishing also declined. As a result, large amounts of dried frozen fish were imported. In 1970 such imports were 18 percent of the total; they had tripled since 1966. The situation grew even worse due to the deliberate policies of the regime; a prohibition on the use of canoes was a blow to small-scale fishing. Paradoxically, the export of fish increased. Modern equipment and wide-reaching fishing privileges benefitted the Soviet Union. In 1971 Moscow negotiated an agreement permitting it to use the republic's territorial waters. By 1979, the fleet's harvest was worth several million dollars.

REHABILITATION: FINANCE AND AID

The rule of Macias Nguema left Equatorial Guinea "in a shambles—economically, politically and socially," according to one U.N. observer.[13] Four years after the dictator's demise, Equatorial Guinea was the poorest country in Central Africa in terms of per capita gross national product (GNP); only Chad came below Equatorial Guinea in this respect—$110 in Chad versus $180 in Equatorial Guinea. Zaire was slightly higher with $210 per capital per annum. The republic's old agricultural rivals, São Tomé and Príncipe, had $370 per annum and neighboring Cameroon had $880 per annum. Most striking was the fact that Gabon, the country that almost surrounds Equatorial Guinea, had a per capital GNP rate of $3,740.[14]

Under Obiang Nguema Equatorial Guinea has entered a period of *rehabilitación*. In the economic sphere the removal of the Macias Nguema government has not resulted in quick gains. Indeed, the period since 1979 has been marked by economic stasis or, in some cases, further decline. For instance, in the first year of the new government cocoa production was only 5,408 metric tons compared with 38,000 metric tons in 1968–1969.[15] Seven years later the U.S. Department of State observed that "there is little industry, and the local market for industrial products is small. The government seeks to expand the role of free enterprise and to promote foreign investment but has had little success in creating an atmosphere conducive to investor interest."[16]

Paying Off Foreign Debts

Until the mid-1980s, the black market and foreign currency traffic were severe problems in Equatorial Guinea. In 1981 Equatorial Guinea imported approximately $58 million in merchandise versus $26 million in exports. The country thus had a $32 million deficit for that year alone. In 1983 the foreign debt was $150 million and annual payments in 1984 were $17 million. Total external debt reached $151.6 million in 1986.[17] Ninety-three percent of this was long-term public debt. Of this, 72 percent was owed to bilateral creditors. According to various indices, Equatorial Guinea was one of the most debt-ridden countries in the world with approximately $420 worth of foreign debt per inhabitant. Since independence the gross national product has decreased about 6 percent per year.

By the end of 1981 Equatorial Guinea's financial situation was parlous. After 1979 Equatorial Guinea had persisted with its own national currency, the *bikwele*. In 1980 a team of Spanish advisers and the International Monetary Fund helped to prepare a budget for the largely bankrupt state. The budget provided for the expenditure of 2,025 million bikwele, two-thirds of which was paid out in functionaries' salaries. The government received 1,651 million bikwele in income, derived mainly from import and export taxes. In 1983 two bikwele were in theory worth one Spanish peseta. In reality, on the black market, the rate was eleven bikwele to one peseta.

In the early 1980s, Equatorial Guinea joined the Union Douanière et Economique de l'Afrique Centrale (UDEAC). In August 1984 Equatorial Guinea entered the Banque des Etats d'Afrique Centrale (BEAC). In January 1985 the CFA franc replaced the national currency and the Bank of Central African States opened a branch in Malabo. Entrance into the CFA zone has had certain ameliorative effects. The country shares its currency with its Francophone neighbors, including Cameroon and Gabon, which eases trade contacts. To some Guineanos this presages political as well as economic integration.

Participation in the CFA zone has its problems. In April 1986 Equatorial Guinea experienced, for the first time, the tenth devaluation of the CFA, which is pegged to the French franc. All purchases in Switzerland, West Germany, and the United States cost 6 percent more. The CFA zone and the UDEAC have opened the country to two types of foreign capital. *Speculative* capital has arrived from neighboring countries. It is invested in everything from fairly large-scale establishments to petty trading. Investors from African countries sell non-African manufactures and most of their profits are expatriated. Such investment finds its opportunity in the undercapitalization of Equatoguineans. The second type of capital is *strategic* capital. Most of it is French and is found in very specific sectors: telecommunications, air transport, airports, and banks.

Competition from the other countries of the CFA-UDEAC zone may work to the detriment of Equatorial Guinea. Each member works to attract investment and to export the same type of products. The inconveniences of having a common and limitlessly convertible currency may not be a blessing; Equatorial Guinea is competing with larger countries that have much greater output.

Rising debt is also a serious problem. In 1987 the U.N. Development Program reviewed the situation:

> The Gross National Product . . . increased, between 1980 and 1986, at an average rate of not more than 3.5 percent, although it should be noted that marked fluctuations were produced within the period. For 1986 a small growth in the GNP is calculated, derived from the low production of cocoa, in some means compensated for by the expansion of forestry activity and the continued flow of international assistance.
>
> The external public debt rises to $123 million, that is to say, 1.5 times the GNP, and its servicing constitutes a heavy burden for the balance of payments, which from 1980 on presents an important deficit. In 1986, service of the debt represented 57% of exports, in spite of the favorable renegotiation which was achieved in 1985 [with] the Paris Club [i.e., aid donors] and with the principal creditors.[18]

By the middle of the 1980s, the total foreign debt reached $150 million: People's Republic of China—$50 million; Spain—$45 million; African Development Bank—$10 million; Arab Bank for African Development—$5 million; Argentina—$5 million; others (e.g., Soviet Union, France, Italy)—$35 million.[19] In 1986 Spain provided 33 percent of the country's external aid. The United Nations provided 27 percent and the European Economic Community 5 percent. The African Development Bank and China each provided 7 percent.[20] The following year the European Economic Community was the principal source of aid (ECU [European Community Unit] 12 million); Spain also gave $14 million and the U.N. Development Program (UNDP) gave $1 million for education. There were numerous other donors; for instance, U.S. development aid for fiscal year 1988 was $850,000; it was $500,000 the previous year.[21]

The cost of rehabilitation is high and both the government and the International Monetary Fund (IMF) have seen the need to reschedule debt servicing. The World Bank estimated the total debt service at $17.2 million for 1988. It will rise to $21.9 million in 1991. Equatorial Guinea should be able to pay the interest, but the fluctuating market for its exports makes repayment of the principal almost impossible. An IMF mission visited Malabo in December 1987 and proffered an arrangement whereby Equatorial Guinea might ease its debt burden. As one of the 62 poorest countries, Equatorial Guinea qualified for the IMF "enhanced structural adjustment facility" (ESAF). This is aimed at helping countries whose economies are particularly subject to fluctuations in world commodity prices. A country's level of borrowing will depend on its balance of payments and the interest rate will be only 0.5 percent per annum. In 1986 the donors' group, the Club of Paris, rescheduled the $150 million in remaining external debt. In December 1988 the IMF approved a three-year structural adjustment facility. The following March, the Club of Paris discussed Equatorial Guinea's debts through 1988 without reaching any firm decisions. Malabo's total external debt was approximately $200 million, seven times larger than estimated export earnings.

Foreign Aid Rehabilitation

In the early 1980s the UNDP drew up rehabilitation plans that coordinated the activites of a number of donors, among them Spain, France, the European Economic Community, the World Bank, UDEAC, BEAC, Gabon, and Cameroon. In April 1982 a donors' round table (International Conference of Donors for the Economic Reactivation and Development of the Republic of Equatorial Guinea) was held. A second development program covered the period 1983–1987. The total value of foreign assistance grew from approximately $25 million for the year 1983 to $35 million in 1986. Thus, the country experienced an important influx of external aid—around $30 million per annum. Unfortunately, due to both deficiencies in the execution of the projects and restrictions in the absorptive capacity of the public administration, much that the plan sought to accomplish was not achieved. A second donor's round table completed its deliberations in November 1988 and was optimistic about solving these problems in the future.

Obiang Nguema visited Europe in November 1980 and received promises of financial support from the European Economic Community and the International Monetary Fund. The latter provided the republic with a loan of $23 million. Obiang Nguema also signed assistance agreements with Belgium, West Germany, the Netherlands, and the United States. In the latter country, a $1 million aid package was proposed in response to the new regime's apparent shift away from the Soviet Union.

Planning mechanisms were in chaos when Obiang Nguema took power. U.N. assistance to the Ministry of Planning and Economic

Development was begun in 1982. Of particular importance was the creation of a Central Projects Unit, which aimed at the preparation of the basic documents for negotiation with donor countries and international bodies. After five years of existence, both the U.N. Development Program and Equatorial Guinea felt that further aid was necessary in order to assure the implementation of the agreements reached in planning, public investment, and international technical cooperation. A 1987 project sought to answer basic questions about the political economy. This project involved consolidating the national planning system and establishing planning units in the principal technical ministeries with the final aim of creating institutional mechanisms to coordinate planning and budget preparation. The project was scheduled to last thirty-two months and receive a total UNDP contribution of $1.3 million.

The UNDP head met with the minister of planning and economic development in March 1987 to review past programs and to preplan a further three-year plan (1987–1991). For that period the country will receive annual outside aid between $32 million and $36 million. Seventy million dollars in nonreimbursable technical assistance (including the U.N. assistance) is predicted. Half of this will be put into foresty, agriculture, and fishing. Sanitation and health care will receive 30 percent. General matters of development policy and planning will receive 10 to 15 percent. The residuum of 10 percent will be put into transportation, communication, health, and education. The development strategy for 1987–1991 hopes to establish an average GNP growth rate of 5 percent as a minimum goal for the next five years.

UNDP support is part of a wider program in which other multilateral and bilateral donors also participate. For instance, France allotted around $2 million in 1986–1988 in assistance for the training of civil servants in the management of public finances. The World Bank established a credit of more than $3 million for the period 1985–1989. This sum was used to acquire five planning experts in the areas of macroeconomics, agriculture, forestry, mining, and education. These experts, especially those of the World Bank, maintain coordination with the activities of the UNDP. The People's Republic of China is an important source of aid in medical, agricultural, hydroelectric, transportation, and other fields.

The former metropole attempted to step into the vacuum created in 1979. In December 1979, the Spanish monarch visited. Obiang Nguema averred that "the future of Equatorial Guinea does not make sense without Spain." Juan Carlos was told that the former colony demanded of the king and the Spanish people "that they make of Equatorial Guinea the long-desired 'Switzerland of Africa'." In March 1980 a commercial bank, Guinextebank, was founded with the support of the Banco Exterior de España. Obiang Nguema promised to return all confiscated Spanish properties. This hope had been kept alive by the Comunidad de Españoles con Intereses en Africa (CEIA), an intransigent group of former plantation owners who have consistently lobbied their government.

In 1979, Madrid airlifted equipment and technicians to the traumatized former colony. Spain pledged $19 million in aid, 500 technicians, and the Spanish Foreign Legion. Equatorial Guinea warmly welcomed these pledges and asked for a five-year period of Spanish administration over social and economic services. Spain supplied $23 million worth of credits. Government administration was revamped by a force of 120 Spanish advisers and many students were sent to Spain. By early 1981, Madrid had sent nearly 3,000 civilian personnel and a group of military advisers, which raised the Spanish presence to almost half of its preindependence level. The Spanish also conducted a search for minerals, especially uranium, through the Instituto Geológico de Madrid.

The relationship has not been without difficulties. Articles critical of the new regime brought forth a ban on Spanish publications from August to October 1981. Early in 1982 a $23 million three-year plan of assistance was rejected because it was conditional on the acceptance of three Spanish consultants in the ministries of finance and budget. Later Spain's Banco Exterior froze its share of the assets of Equatorial Guinea's Central Bank. Some reports indicate that Spain spent $50 million in Equatorial Guinea in 1979 and that a further $80 million had been requested. Spanish Chamber of Deputies figures, however, indicate that the actual aid from Spain was only $20 million in two years. Most of it was spent on Spanish imports and technicians. In January 1989 Spain agreed to make concessions on the Equatorial Guinean debt. About a third of the amount owed Spain, 1.8 million pesetas ($15.6 million), was written off. The remaining debt is repayable over fourteen years and has an eight-year grace period, subject to the approval of other aid donors.

There were difficulties in the summer of 1985 when Iberia airlines suspended flights to Malabo because of supposedly inadequate safety precautions. Late in 1985 the director of the Spanish Institute of Cooperation arrived to review the situation. Afterward the Spanish continued to provide a budget subsidy and Obiang Nguema continued to profess adherence to a special relationship with Spain.

Spain has had to struggle to retain its preeminent position. After 1979, France approached gave Fr11 million (approximately $2.5 million) for Malabo port facilities, geological prospecting, fishing surveys, and university scholarships. The French cooperation minister visited immediately after the 1979 coup and pledged to broaden collaboration. Obiang Nguema visited France the following month. These developments were not unnoticed in Spain. In 1982, the French foreign minister sought to assuage Spanish suspicions regarding French intentions. However, the jostling between Spain and France increased in the 1980s with renewed projects for exploiting oil. In 1984 France became the chief supplier of imports for the first time. However, Spain regained its preeminence the next year when it supplied 30 percent of the goods coming into Equatorial Guinea.

Increasingly it seems that Equatorial Guinea is an extension of Francophone equatorial Africa. In this change of orientation, "Spain has little option but to accept the *fait accompli* of Equatorial Guinea moving into a predominantly French sphere of interest as a member of . . . the Franc zone. Though Spain might lose some prestige, it could be quite happy to be rid of the responsibility for bailing out the bankrupt Equatorial Guinea economy." Furthermore, "the USA, for its part, favours membership in the CFA zone, as it believes that France would be better placed than Spain to discourage Equatorial Guinea from resorting to aid from the Soviet bloc."[22]

By the middle of the 1980s, in spite of the continuing predominance of aid from Madrid, Spanish advisers were losing influence to an influx of French technical exports. In June 1985 Obiang Nguema visited Paris and met French President François Mitterrand. The African president signed agreements for French technicians, the granting of fifty scholarships, and the extension of French aid for food production, energy, and resource development projects. In April 1986, the French government reorganized its aid program and attempted to draw closer to the former colonies of Spain and Portugal. The reorganization guaranteed closer contact with the minister of cooperation in the French Ministry of Foreign Affairs. In 1986 the French–Equatorial Guinean Joint Commission began work and the Franco–Equatorial Guinean Friendship Committee was formed. In the same year a report prepared by a French expert recommended that the study of French be obligatory at certain levels in the Equatoguinean schools.[23] In 1987 a French company, France Cable, was given the contract to install a new telecommunications system. The former colony had rejected a proposal from the Compañia Telefónica Española. The French Caisse Centrale de Coopération Economique (CCE) is also cofinancing plans to revive the national airline and construct a hydroelectric station. In September 1988 Obiang Nguema visited France again and asked for extensive aid.

At present Equatorial Guinea is engaged in a balancing act between Spanish and French capital. In 1987 Obiang Nguema noted that the Spanish-backed Guinextebank had functioned very badly but thought it important to maintain links with Madrid. Despite this, in January 1988 the Banco Exterior de España announced its withdrawal from

Guinextebank, in which it had a half interest. Equatorial Guinea bought the Banco Exterior's share and Madrid provided $12.7 million to cover the bank's debts. This did not cover its debts to BEAC, however, which functions as a central bank. In February 1989, Spain agreed to encourage an exclusively private bank to overcome the difficulties caused by the withdrawal of Guinextebank.

For a period, the government also backed the Banco de Crédito y Desarrollo (BCD), in spite of faltering confidence. In 1987 the IMF called for the elimination of the BCD; Malabo agreed in 1988. The Obiang Nguema regime also appears wary of the power of yet another French-backed bank, the Banque Internationale de l'Afrique Occidental—Guinea Ecuatorial (BIAO-GE). Fifty-one percent is owned by France and it has been accused of providing insufficient service. In 1987 BIAO-GE opened an agency in Bata, but this did not mute continued criticism. At the end of 1989 BEAC agreed to refinance medium-term credits given to BIAO-GE.

Structural adjustments have been suggested to make Equatorial Guinea more competitive with its much larger neighbors. One is the creation of a new national bank. The failure of Guinextebank and the Banco de Crédito y Desarrollo does not augur well for this approach. An Equatorial Guinean economist, Marcos-Manuel Ndongo, has nevertheless argued that an appreciable increase in money supply is needed. To maintain Equatorial Guinea's economic independence, he urges a national bank. Equatorial Guinea could partially retire from the CFA-UDEAC zone, perhaps retaining observer status. It could then open negotiations with a country or group of interested countries to establish a new currency. Negotiations would also encourage private investment and link Equatorial Guinea to preferred treatment in the markets of its loan partners. Equatorial Guinea and its creditors would negotiate a fixed parity between the hard currency and the new national currency (for example, one to three).[24] These proposals are similar to those made by Garcia-Trevijano at the time of independence. Unfortunately, Equatorial Guinea is too small an economy to sustain its own national currency. At the same time, the repubic risks becoming an economic appendage of its neighbors by using the CFA franc.

In 1969 Spain proposed to guarantee the value of the *peseta guineana*. Today, after three changes in the monetary system, Hispano-Guinean financial cooperation continues. In September 1988, the Oficina de Cooperación con Guinea Ecuatorial (OCGE) announced a slight augmentation in assistance, in spite of such past ventures as Guinextebank. Fernando Riquelme, the head of OCGE, assured officials of a commission of the Cortes that past irregularities has been overcome.

EXPORTS AND INFRASTRUCTURE

In the mid-1980s the country's exports were valued at $24.8 million. Today major products are cocoa, coffee, and lumber; Spain remains a major market. In addition, some exports reach France, Germany, Italy, and the Netherlands. Those industries that have developed have not been labor intensive. There is some manufacture of paper paste, flour, alcohol, prepared fish, and soap. In 1980 a baby food company was started; the venture is financed 50 percent by Spain, 25 percent by Afriexport, and 25 percent by private capital.

Imports far outstrip exports; in 1985 they amounted to $34 million. Heading the list of imports are foodstuffs, beverages, textiles, machinery, motor transport, agricultural products, fuel, and tobacco. The republic's chief suppliers are Spain, France, Italy, the Netherlands, the EEC, the USSR, the People's Republic of China, West Germany, and the United Kingdom.

Petroleum

France and Spain have been among those countries whose investors have been seeking petroleum, which is Equatorial Guinea's most enticing resource. Perhaps as recompense for the solicitude of the former metropole, the new government granted exploration rights to the Spanish state-owned company, Hispanoil. The Spanish company formed a joint-venture with Equatorial Guinea, Empresa Guineano–Española de Petroleos, in 1980. The petroleum company was granted an exploration lease to a 2,072-square-kilometer area north of Bioko. Plans called for the installation of a rig by the end of 1982 and the beginning of production in 1983. Regrettably, to date production costs have not made Equatorial Guinean petroleum worth exploiting.

Significantly, the French companies, Elf-Aquitaine and Compagnie Française des Petroles, have also received exploration rights. Economic penetration takes an interesting form. In Rio Muni the oil development agreement gives Elf-Aquitaine 40 percent of the profits and the government of Gabon 30 percent. The government in Malabo also receives 30 percent. In terms of gasoline consumption and distribution within the country, French interests also achieved the upper hand. "Total," a French company, received the right to distribute gasoline throughout the country. This function had previously been exercised by the Empresa Nacional Petrolifera. By the mid-1980s the division of concessions in Equatorial Guinea was complete. French interests held more than half of the available onshore and offshore lands. Elf-Aquitaine and its affiliates held 2,232 square kilometers; Getty Oil held 1,968 square kilometers, while

Hispanoil had 1,972 square kilometers. In addition to its distribution rights, "Total" had 5,594 square kilometers of oil lands. In 1984 gas was discovered off the northern coast of Bioko. In late 1989, a British-based company, Clarian Petroleum, obtained 198,000 hectares near the island of Corisco.

Agriculture

Foreign assistance to the agricultural sector, especially for cash crops, has been forthcoming. This serves to obscure the fact that the majority of Equatorial Guineans are subsistence farmers. The main food crop on Bioko, malanga, grows in abundance. The problem in the late 1980s was to provide for its transport to areas of relative need. In 1984 the country estimatedly produced about 54,000 metric tons of cassava, 35,000 metric tons of sweet potatoes, 18,000 metric tons of bananas, 8,000 metric tons of coconuts, 5,200 metric tons of palm oil, and 3,000 metric tons of palm kernels. Although the country could be self-sufficient in food production, food imports, including tobacco and beverages, were 25 percent of total imports in 1983. In 1986 the percentage has risen to 47 percent.

Cocoa represents a past success that some experts would like to revive. Only 15,000 hectares of cocoa were in production in 1982 and the quality was declining due to neglect. Less than one-tenth of the owners of cocoa lands have returned to farm.[25] In the late 1980s this continued to be true. In 1987 some 20,000 hectares were still abandoned and another 20,000 hectares were being worked by farmers hoping to secure a firm title after five years of working the land.

The African Development Bank has started a project intended to reintroduce cocoa growing to some 5,500 hectares. Only 700 hectares were put into cultivation in the first four years of the project. The Organization of Petroleum Producing Countries gave a loan of $1.5 million in 1984 for agricultural development; the following year the Arab Economic Development Bank contributed $2.8 million. In 1986 the World Bank approved a $10.6 million import credit to assist Equatorial Guinea in financing necessary inputs in agricultural production. The World Bank project is cofinanced by the European Development Fund and scheduled to run through 1990.[26] The Arab Economic Development Bank is col-laborating in the World Bank program; the cost of their combined effort is $16.2 million. Also, with the aid of the World Bank, the Spanish firms of Apra and Mallo have reentered cocoa and coffee marketing. A similar function is provided by the French Societé Planteurs, SA, and its subsidiaries Sogec and Sodis-Guinea.

The central economic problem continues to be labor shortage. In 1980 the government took a rural census. Shortly thereafter, it issued identification passes in an effort to control labor availability. In 1981 and 1982 the government attempted to persuade Nigerian migrants to return. Some did, but chiefly as petty retail traders. Small numbers of Ghanaian laborers arrived and the government was reportedly looking to Rwanda, Burundi, and Burkina Faso, as well. The wages paid on the island are an obvious disincentive. The United Nations recommended a daily salary of 14,050 bikwele in 1984; at the time the monthly salary was 8,500.[27] Wage rates in the late 1980s were still not enough to attract workers. Agricultural labor was paid between 1,200 and 1,500 CFA francs per day in Cameroon and Côte d'Ivoire; monthly wages were between 30,000 and 39,000 CFA francs. On Bioko monthly wages were between 11,000 and 15,000 CFA francs.[28]

Labor negotiations play a significant part in the tangled relations between Malabo and Lagos. In March 1985 the foreign minister and the minister of defense paid separate visits to Nigeria and found that Lagos insisted on guarantees of good treatment for its nationals. Moreover, the issue of labor abuse is a continuing problem. In early 1985 a Nigerian migrant was slain by the police. Nigeria sent naval vessels and an aircraft. The larger country demanded compensation, an end to the illegal labor smuggling, and the right of workers to appeal to their consulate in Malabo. Equatorial Guinea acquiesced, but this action did not gain the desired labor flow.

Labor shortage, in combination with market factors, has had its impact. Since 1979 production has varied between 5,000 and 10,000 metric tons, well below the 38,300 metric tons produced in 1968–1969 (see Table 4.1).[29] Cocoa earnings may have reached their peak and they are not expected to rise because of the policies of the International Cocoa Organization and world inventory levels.[30]

In the mid-1980s, out of approximately 50,000 hectares planted with cocoa, only 20,000 hectares, mostly on Bioko, were in production. In late 1987 approxiamtely 6,300 people were engaged in cocoa cultivation. Given this level of manpower investment, production is not expected to go much beyond 7,000 to 8,000 tons per annum.[31] By 1988 the government had adopted tax policies aimed at facilitating production. Also under the Lomé Convention, a fund was set up to stabilize export earnings. For example, in 1988 funds were delivered to compensate for low cocoa earnings in the previous year. Serious problems remain; among them are the destruction of at least 20 pecent of Bioko's crop by squirrels, the underpayment of export taxes by foreign companies, a quality discount of £60 per ton in the importing countries, and the high price of pesticides.

In the spring of 1988 the government announced the creation of a cocoa marketing board, Inprocao. Italian interests promised to put an estimated $10 million into the project. They were reportedly also con-

TABLE 4.1
Cocoa Production, 1968–1988

	Amount Produced (Tons)	Percentage of World Market
1968–1969	38,000	3.06
1969–1970	24,000	1.67
1970–1971	30,000	2.00
1971–1972	22,000	1.39
1972–1973	10,000	0.72
1973–1974	12,000	0.83
1974–1975	13,000	0.84
1975–1976	10,000	0.66
1976–1977	6,000	0.45
1977–1978	7,000	0.45
1978–1979	8,000	0.54
1979–1980	5,000	0.31
1980–1981	8,000	0.48
1981–1982	8,000	0.46
1982–1983	10,000	0.65
1983–1984	8,000	0.53
1984–1985	8,000	0.43
1985–1986	5,000	0.42
1986–1987	5,000	—
1987–1988	7,000	—

Sources: Data for 1968–1969 to 1985–1986 from Cord Jakobeit, "Aquatorialguinea: Schwierige Rehabilitation," in Afrika Spektrum (Hamburg: Institut für Afrika-Kunde, no. 2, 1987), p. 150; data for 1986–1987 and 1987–1988 from Economist Intelligence Unit, Gabon, Equatorial Guinea. Country Profile, Annual Survey of Political and Economic Background (London: Economist Intelligence Unit, 1988–1989), p. 38.

sidering the development of 10,000 hectares. The majority of this land would be *fincas* abandoned by the Spaniards. The stated aim of Inprocao is to stabilize prices and to provide credit to farmers. The marketing organization will assume the functions of the Cámara Agrícola and design plans for cocoa expansion.

Smallholders may be the future of Bioko. They now produce the majority of the cocoa crop. This is, in part, an outgrowth of land policy. In 1984 the government seized lands that had been abandoned since the exodus of European planters. Some of the lands were leased to small producers. Smallholders and their families provide most of their own labor needs. There are also "permanent laborers" or hired nonmigrant farm workers. There were not more than 6,300 in 1986. At the same time, foreign laborers, mainly Ghanaians, were estimated at only a few hundred. Another category of laborer is the *parcelista*, a tenant farmer who cultivates up to 5 hectares owned by someone else. The landowner provides materials and gives a cash advance and in return the *parcelista* agrees to deliver all of the crop to the landowner. If the crop's value exceeds the value of the advance, the tenant receives the difference. If this is not the case, the tenant is bound to work the owner's land the following year.

Parcelista tenancy is increasing in the face of continuing labor shortage. If the government does not push land reform, the system will exacerbate existing problems. For example, "today's main problems refer to the increasing indebtedness of many parcelistas who can be easily cheated by the estate owners, thereby creating a system of landless and poor laborers who are not particularly interested anymore in expanding cocoa production since there are no chances whatsoever to have access to an independent status." Also, "the system is not tuned to encourage endeavors such as replanting or rehabilitation which do not give a fairly immediate return for effort expended."[32]

Before independence there were over 30 cooperatives in Equatorial Guinea. Early in 1980 the government began a program to reestablish them. By 1984 there were 29 cocoa cooperatives on Bioko and 18 cocoa and coffee cooperatives on the mainland. Under a contract from the Agency for International Development, the U.S.-based National Cooperative Business Association (CLUSA) entered the country in 1980. By 1987 it had coordinators working on Bioko and in Rio Muni. CLUSA provides training in accounting and the running of cooperative stores. The organization emphasizes grassroots organization and encourages the populace to organize committees (*comités de gestion*). CLUSA also helped organized consumer stores (*enconomatos cooperativos*). Men and women's cooperatives are organized separately. The former concentrate on cash crops and CLUSA has devoted much effort to providing transport. Women's cooperatives concentrate on the production of subsistence crops, sometimes with considerable success.[33]

Efforts have been made to revive the cultivation of other crops. In October 1985 the International Agricultural Development Fund announced that it would put $2.4 million forward in three rural development projects for coffee, peanuts, and manioc in northeast Rio Muni. The decline of coffee export occurred for the same reasons as the decline of cocoa—dearth of labor, lack of bookkeeping, and semigovernmental expropriation of lands. The International Coffee Organization has sought to ease the country's situation. For example, Equatorial Guinea received an allotment of 20,706 bags of 60 kilograms each (approximately 1,242 tons) in 1985–1986. This action was designed to increase the republic's holdings of foreign exchange.

Forestry

Forestry is a leading source of foreign earnings. The U.N. Food and Agriculture Organization says that the country could produce 300,000

TABLE 4.2
Composition of Exports, 1970–1987 (in percentages)

	Cocoa	Coffee	Wood	Other	Total
1970	66	24	9	1	100
1976	87	10	3	0	100
1977	82	8	10	0	100
1978	98	0	2	0	100
1979	96	0	3	0	100
1980	85	0	15	0	100
1981	81	0	19	0	100
1982	72	3	25	0	100
1983	60	0	40	0	100
1984	49	5	51	0	100
1985	69	5	26	0	100
1986	46	9	45	0	100

Source: Cord Jakobeit, "Aquatorialguinea: Schwierige Rehabilitation," in Afrika Spectrum (Hamburg: Institut für Afrika-Kunde, no. 2, 1987), p. 151. Reprinted by permission of the Institute für Afrika-Kunde.

cubic meters per annum. The volume of wood exported was 138,000 cubic meters in 1986 (a 55 percent increase over the year before). This, combined with an increase in price, raised forestry earnings to $6.5 million.

Wood amounted to approximately 47 percent of export earnings in 1986 and approximately 15 percent of government revenues. (See Table 4.2.) In spite of this good showing, lumbering still did not reach the 230,000 cubic meters predicted for 1986. Estimates from the IMF for 1987 were 160,000 cubic meters. Numbers for the first six months of 1987 (70,000 cubic meters) were slightly less than the comparable period in 1986. For 1988, approximately 186,000 cubic meters should have been produced; wood accounted for more than half of export earnings. For 1989 timber exports were projected at 194,000 cubic meters.[34]

Prices have risen since 1985. Demand has been especially strong from Japan, the world's largest importer, as well as from South Korea, China, and Taiwan. Another factor in the surge in forestry activity has been an upswing in European furniture and construction industries. Prices promise to rise; the founding of the International Organization for Tropical Woods in 1987 should limit violent price fluctuations.

Okume and akoga wood constitute the bulk of Equatorial Guinea's lumber exports. This poses a problem for the future because the world market for okume is saturated. In addition, export duties do not adequately discriminate between the value of different wood grades; this has operated as an export disincentive. The government has prepared a new and more complex system of export duties and is training personnel to implement it. A forestry marketing body was set up in Bata in 1985.

A new organization, funded by French aid, has now been proposed to supplement it.

Initial forestry concessions last for five years, but this period is probably too short for the full amortization of investments. Concessions are renewable, however. The Onassis-Roussel group gained a large concession in 1985 in return for a promise to build a road between Kogo and Evinayong. Also the Italian firm, Siem, in collaboration with the government, is foresting 50,000 hectares and producing 20,000 cubic meters of rolled wood per annum. At present fourteen companies are active in silviculture. Eight are Spanish, four are Equatoguinean, and two are French. The Semge company, which is French, has a very large concession of 80,000 hectares. Like the other firms, it has experienced the difficulties of doing business in Equatorial Guinea. All companies have long delays in receiving payments for exports. Because the confused banking situation makes it hard to raise funds within the country, these companies are forced to seek capital in Europe and to depend on overdrafts.

In July 1987 the EEC signed a third aid agreement with Malabo: Eighty percent of the program's resources will go into forestry. In 1988 there were only 1,200 people employed in silviculture.[35] Sawing and rolling timber are the only ancillary activities at present. The government plans to increase the percentage of processed wood to logs in future exports. At present, the country has only enough sawmilling capacity to process 40 percent of the trees cut.

Deforestation is proceeding swiftly. Approximately 2,500 hectares are cleared annually without any replanting. About 3,000 hectares per year are cleared for cultivation of food crops. "The exploitation of the forest is," according to the UNDP, "executed in a manner predominantly empirical, improvised and inefficient, with the consequent danger of destruction of this resource in the long term."[36] The program has urged ecological studies, an inventory, and a map of forestry resources. It also plans to bring in "the minimum human and material infrastructure in the forestry administration." The program will last four years and take in a UNDP contribution of $975,000. During the period 1987–1991 a number of projects, totaling $5.7 million in aid money, will tackle problems connected with lumbering, including reforestation. To increase government income from exports, 60 percent of forestry exports will consist of processed wood in 1990. New concessions will demand more processing and duties on unprocessed wood will be progressively increased.

Although silviculture is surging ahead, it is constrained by the lack of infrastructure. Timber exports were 67,792 cubic meters in 1982;

Logging camp, Rio Muni. (Photo courtesy the United Nations)

69,522 cubic meters in 1983; 86,041 cubic meters in 1984; 89,464 cubic meters in 1985; 137,963 cubic meters in 1986; and 160,000 cubic meters in 1987.[37] During 1986–1989 lumbering fell well below the 200,000 cubic meters set as a goal in agreement with foreign forestry companies. The export of lumber from Bata is held up by harbor congestion. It takes fourteen days to load 2,000 cubic meters of wood. If harbor improvements are carried out, it will take only four days. A plan for these improvements is being funded by $4.8 million of Italian aid money. The Ente Autonomo del Porto di Trieste controls 60 percent of Promoport Guinea, the Bata port authority. Promoport has a fifteen-year contract. Early projections on the amount of wood capable of being shipped were far too optimistic. The road network, which connects with the ports, is also in a state of disrepair. In addition, many roads are too small to accommodate the trucks used by forestry companies. In some parts of the country there are still no roads.

Mining, Fishing, and Ranching

Rio Muni has appreciable subsoil wealth. The territory contains copper, uranium, and iron ore. Equatorial Guinea passed a mining law

Fishing boats on the beach at Annobón. (Photo courtesy Frank Ruddy)

in 1981 and the French Bureau de Recherches Géologiques et Minières
(BRGM) has examined the potential for mineral exploitation. There are
possible manganese and tantalum reserves. In 1986 a mixed Spanish–
Equatorial Guinea company, Guineana Española de Minas, SA (GEMSA),
conducted laboratory and aerial studies.

Under Obiang Nguema efforts have been made to rehabilitate and
redirect fishing. In 1982 the national catch was 2,500 metric tons. A
1981 agreement with Nigeria gave Equatorial Guinea the right to fish
in Nigeria's territorial waters and vice versa. Spanish fishing interests
were also invited into the country. The Afripesca and Ebana companies
built new refrigerated warehouses for imported fish, but the government
complained about Spanish delay in providing fishing boats to replace
Soviet boats. After 1984, the European Economic Community (EEC) tied
its economic aid to permission for community vessels to fish in Equatorial
Guinean waters. In 1986 Equatorial Guinea and the Community signed
an agreement to regulate EEC fishing. The Pechaud Guinée Equatoriale,
an affiliate of the French Pechaud et Compagnie Internationale, carries
out much of the republic's fishing operations. The U.N. Development
Program has conducted studies on fisheries. Approximately 8,000 tons
of fish should have been produced by 1990. This would be enough to
cover internal demand and is especially important for protein deficient
diets in Equatorial Guinea.

Cattle and poultry production have received some aid. In the past twenty years ranching was largely abandoned and now virtually all meat products are imported. As mentioned, South Africans have been somewhat ominously involved in revitalizing cattle raising on Bioko. Spain, the United States, and the African Development Bank have provided technical and financial aid for poultry production.

Transportation Links and Power Sources

Transportation to Equatorial Guinea and between its parts has been improved. After 1979 the government requested the continuation of Transmediteránea's service. The company declined because the proffered government subsidy was insufficient. Three years into the new regime, the Compañia Guineana de Navegación Maritima, SA, was formed with 51 precent state participation. The company was formed in association with the Belgian Societé Belge de Consignation et d'Affrètement Saint Nicolas and permitted to establish a link between the republic and Anvers, Rotterdam, and Hamburg. In 1985 maritime linkages were further improved when the French company Delmas-Vieljeux linked Bata and Malabo to French ports. Private initiative has not been enough. The EEC has aided in regional maritime cooperation through the creation of a sea linkage between Equatorial Guinea, São Tomé and Príncipe, Cameroon, and Gabon. In January 1990, a French concern, Chantiers Modernes, received a contract to expand the Bata airport with the aim of making it the chief entry point for the country.

In 1985 a group of French companies already present in Gabon and Cameroon created a construction firm in Equatorial Guinea. Other construction companies, mainly Spanish and Italian, continue to operate. Construction work on a major hydroelectric power station at Riabba on Bioko cost $37 million. It was opened on August 1, 1988. Major funding was pledged by the European Development Fund and the Caisse Centrale de Coopération Economique (CCCE). The regular provision of electricity to the capital is finally assured.

AN OVERVIEW

The economic situation since 1979 reflects halting development in spite of high inputs of foreign aid. In 1984 the IMF identified the country's problems as a lack of indigenous experts, an overvalued currency, seriously diminished productive capacity, and dependency on external production inputs. Entrance into the CFA franc zone devalued the currency by no less than 85 percent. The country has also followed a strategy of reduced expenditures and property privatization. In spite of much planning and an envisaged growth rate of 4 percent per annum, the economy has not gone as planned. The need for production inputs— for example, copper sulphates in cocoa growing—has suffered because of higher import costs following devaluation. This has increased the debt rate and generated a liquidity crisis. Additional World Bank import-credits have become necessary because of massive capital flight following the introduction of a convertible currency.

Lumber and cocoa production have been targeted as areas of investment for the 1990s. The fluctuating markets for these products have not put the economy on a firm basis. Indeed, increased cocoa production in 1987–1988, during a period of falling prices, showed the fragility of the strategy of reviving plantation agriculture. A recent economic observer noted that "in the long run the success of this new policy [i.e., forestry and cocoa production] and the stability of the Obiang regime will require both unrestricted access to further credits and enlarged participation of the entire, still powerless population." If not, "all undertaken adjustment measures will only bring medium-term results."[38]

NOTES

1. Max Liniger-Goumaz, "La république de Guinée Equatoriale. Une indépendence à refaire," *Afrique Contemporaine* 105 (September–October 1979), p. 18.

2. Economist Intelligence Unit, *Congo, Gabon, Equatorial Guinea. Country Report. Analyses of Economic and Political Trends Every Quarter* (London: Economist Intelligence Unit, 1988), no. 3, p. 31.

3. *African Contemporary Record, 1971*, ed. Colin Legum (London: Rex Collins, 1972), p. B505.

4. Suzanne Cronjé, *Equatorial Guinea, The Forgotten Dictatorship: Forced Labour and Political Murder in Central Africa* (London: Anti-slavery Society, 1972) p. 31, citing addendum to agreement of Commerce and Payments, May 19, 1969.

5. *African Contemporary Record, 1968–1969*, p. B457–B458.

6. *African Contemporary Record, 1971–1972*, p. B505.

7. Cronjé, *Equatorial Guinea*, p. 32.

8. *Ibid.*, citing *Marchés Tropicaux*, May 2, 1975, p. 1295.

9. *Ibid.*, p. 33, citing the U.N. Development Program, "Assistance Requested by the Government of Equatorial Guinea for the period 1974–1978." Country and Intercountry Programming. Governing Council, 19th sess., January 15–31, 1975. Item 4, DP/GC/EQG/R.1, October 23, 1974, p. 20.

10. Max Liniger-Goumaz, *Statistics of Nguemist Equatorial Guinea* (Geneva: Editions du Temps, 1986), p. 37.

11. *West Africa*, October 6, 1975, p. 1198.

12. René Pélissier, "Autopsy of a Miracle," *Africa Report* 25, 3 (May–June 1980), p. 11. According to Artucio, the number of workers actually recruited in 1977 was 25,000. He gives the number of dependents as 15,000. Alejandro Artucio, *The Trial of Macias in Equatorial Guinea: The Story of a Dictatorship*

(Geneva: International Commission of Jurists and International University Exchange Fund, 1980), p. 8.

13. "African Update," *Africa Report* 25, 3 (May–June 1980), p. 33

14. Jane Martin, ed., *Global Studies: Africa* (Guilford, Conn.: Duskin Publishing Group, 1985), p. 113, citing *Africa News*, November 7, 1983.

15. República de Guinea Ecuatorial, Presidencia. *Reseña estadística de la República de Guinea Ecuatorial* (Malabo: Secretaria de Estado para el Plan de Desarrollo Económico y Cooperación, Dirección Tecnica de Estadística, 1981), p. 38.

16. U.S. State Department, *Background Notes, Equatorial Guinea*, (Washington, D.C.: U.S. Government Printing Office, 1986), pp. 1, 6.

17. Economist Intelligence Unit, *Quarterly Economic Review of Gabon, Cameroon, CAR, Chad, Equatorial Guinea* (London: Economist Intelligence Unit, 1988), no. 1, p. 25.

18. U.N. Development Program, "Tendencias, estrategias y prioridades de cooperación tecnica." (Planning report for the Third Donor's Conference on Equatorial Guinea, Malabo, 1987), p. 2.

19. *Ibid.*

20. *Courier* (Nigeria) 107 (January–February 1988), p. 34

21. Economist Intelligence Unit, *Quarterly Economic Review*, no. 2, 1988, p. 22.

22. Economist Intelligence Unit, *Quarterly Economic Review*, no. 3, 1983, p. 39.

23. República de Guinea Ecuatorial, Ministerio de Planificación y Desarrollo Económico. Dirección General de Estadística. *Informe general de la estadística educativa para la enseñaña del frances.* Report prepared by Luc Cohen. Malabo, July 1986 [heading: "la educación en cifras"].

24. Economist Intelligence Unit, *Gabon, Equatorial Guinea. Country Profile, Annual Survey of Political and Economic Background* (London: Economist Intelligence Unit, 1988–1989), p. 37.

25. *African Contemporary Record, 1984*, p. B226.

26. Interview with Dr. Robert Kintgaard, World Bank consultant, Malabo, November 8, 1987.

27. Liniger-Goumaz, *Connaître*, p. 152.

28. Cord Jakobeit, "Equatorial Guinea: Long Term Viability of Cocoa Production on Bioko." (Unpublished report for the World Bank Cocoa Rehabilitation Project, Hamburg, 1987), sec. 5, n.p.

29. Liniger-Goumaz, *Statistics*, p. 37; Cord Jackobeit, 'Aquatorialguinea: S chwierge Rehabilitation," in *Afrika Report* (Hamburg: Institut für Afrika-Kunde, no. 2, 1987), p. 150.

30. Economist Intelligence Unit, *Quarterly Economic Review*, no. 3, 1988, p. 31.

31. Jakobeit, "Equatorial Guinea," sec. 5.

32. *Ibid.*, app. 1–3.

33. Interview with J. Colon, director of CLUSA, Malabo, November 8, 1987.

34. Economist Intelligence Unit, *Quarterly Economic Review*, no. 3, 1988, p. 29.

35. *Ibid.*

36. U.N. Development Program, "Tendencias, estrategias y prioridades de cooperación tecnica." (Planning report for the Third Donor's Conference on Equatorial Guinea, Malabo, 1987), p. 12.

37. Economist Intelligence Unit, *Gabon, Equatorial Guinea. Country Profile*, 1988–1989.

38. Jakobeit, "Aquatorialguinea," p. 22.

5

Society and Culture

Present-day Equatorial Guinea presents a contradictory picture. In 1988 a Nigerian journalist found:

> There is no desert without an oasis. Amid the poverty of Malabo, the rich live in style, worthy of their class. There are as many as 20 privately owned Mercedes on this island. Quite a few, too, are owned by the government, which also parades some official Lada cars imported from the Soviet Union. Though the food is costly, the bars are well-stocked with wines, spirits and beer imported from Cameroun, Spain and South Africa. Bioko, being an island, has a large number of foreigners. . . . The consequence of this is that Malabo girls get introduced to adult life quite early.[1]

Sadly, this picture is a great improvement on the one that the city would have presented ten years earlier. An influx of foreign technicians and Moroccan troops give an air of unreality to the capital where one sees more foreigners in motor vehicles than nationals. Malabo now has 50,000 inhabitants and there are a number of uninhabited or partially inhabited dwellings. Prices are extremely high; a thirty-mile roundtrip from Malabo to Luba in an officially approved taxi can cost $100. Although several hotels list themselves as open, only one is in good working order. The capital has only one major restaurant and this often doubles as a site for official social functions. There is no daily newspaper. *Ebano*, published in Spanish, circulates irregularly in the capital. In Bata *Pototpoto* appears on the same basis. The nucleus of a national museum exists in Malabo but is not open at present. Bata has an art museum that contains sculpture and painting, some by the internationally acclaimed artist Leandro Mbomio.

Future museum expansion is promised. The museum in Bata, with the collaboration of Tecnica Española, is planning studies of the flora and fauna. Collaboration will involve the continuation of the scientific study of the Equatoguinean environment; the creation of a functioning

Parade scene in Malabo. (Photo courtesy Dagmar Figl-Mazelis)

museum of natural sciences, the training of a curatorial staff, the creation of protected wildlife areas, adoption of a legal code for wildlife protection, and the incorporation of Equatorial Guinea into organizations dedicated to wildlife preservation.

Many elements of traditional culture remain. This is especially true in Rio Muni where music retains some of its socioreligious functions. The *mvet* is the principal traditional musical instrument among the Fang. It is a type of zither made from a calabash and has a palm stem and strings made from plant fibers. A number of other instruments are also used by the Fang: the *mbeny*—a skin-covered hollow trunk of over one meter in length; the *ngom*—a shorter instrument of the same type; the *mendjang*—the xylophone; the *nlakh*—a horn; the *ngombi* (or *ngoma*)—a harp with up to eight strings.

Some dances and performances remain particular to certain groups. The *ivanga* dances belong to the coastal peoples of Rio Muni. Formerly these were danced exclusively by the Benga, who had received them from the Mpongwe of Gabon. They were done at night and were directed by a queen, or *akaga*, whose role is hereditary. Only women dance; their faces, hands, and legs are painted white and they wear two large white plumes in their hair.

Among the Fang, dances are used to conjure up or exorcise spirits. One of the most popular celebrations is the *abira*, a means through which the Fang attempt to clear away malevolent spirits. A dance, the *balele*, uses violent contortions to this end. Groups of female dancers, the *onxila*, dance in short skirts of vegetable fibers trimmed with strips of monkey and leopard skins. They wear beads, necklaces, metal and fiber bracelets, and colorful plumage of pheasant feathers, which they frenetically shake during the dance. The *onxila* are especially important in Niefang and Micromesang.

There is also music and dance that often includes use of the guitar and incorporates themes derived from outside cultures. Much more recently the music of Equatorial Guinea has been influenced by the "pop" music of Cameroon, which, in turn, is partially inspired by European and especially Afro-American models.

The Fang "palaver house," or *abá*, retains some of its importance. In most villages an area of the palaver house is still cleared and can usually accommodate all of the men. The construction of a palaver house is very simple. Although it is bigger, it is made of the same materials as the other dwellings. A bench of bamboo for those who wish to sit and look out runs the length of the house; most palaver houses lack complete walls. In the center of the structure is a hearth and hunting and war trophies decorate the walls.

RELIGION

As elsewhere, "Westernization"—the imposition of European culture on non-European groups—has had an impact on religion. However, the term should be used guardedly in Equatorial Guinea because of the slow penetration of both the capital and culture of the European colonizer. Elements of Western culture permeated the mainland portion of the colony but only after being transmogrified and incorporated into various forms of African cultural expression. The Fang retained much of their indigenous culture throughout the colonial period. On Bioko this was not the case. The Bubi were subjected to intensive Hispanicization. Mission activity, coupled with the demographic decline, resulted in a population that officially is one of the most Catholicized in Africa

Except for the period of the Second Republic (1931–1936), Spanish colonialism was remarkably free of anticlericalism and relied heavily on the church to achieve its aims. Catholic missionaries, Los Hijos de del Inmaculado Corazón de María (the Claretians), entered the country in 1883. In 1917 they had 80 Spanish and 6 French missionaries in Spanish Guinea. Moreover, there were 30 Conceptionist Sisters; 5 of them were French. A Catholic school had been started at Banapa in the 1860s; later it became the colonial seminary. The first black priest, J. M. Sialo, graduated in 1923.

Wedding procession on Annobón. (Photo courtesy Frank Ruddy)

Spanish Guinea had the highest proportion of priests to population in Africa. In 1959 the apostolic vicariate of Bioko administered 58 priests, including 10 Africans. There were 19 parishes, 28 religious communities, 2 cathedrals, 4 seminaries, 315 chapels, and 16 missionary stations. In 1962 there were 61 priests, 15 of whom were Africans. In addition, there were 17 monks, 23 parishes, 322 chapels, 17 missionary stations, 20 religious schools, 1 clinic, and 1 orphanage directed by Spanish priests. By 1965 proselytizing reached a crescendo as the Francoist regime sought to leave its indelible imprint on black Africa: "Native cults and aboriginal beliefs are harried with all the vigor of the sixteenth century. African customs are suspect and are harnessed or suppressed altogether in favor of the only true values of triumphant Hispanism: love of the Spanish mother country, the Caudillo [i.e., Generalissimo Franco] and the Church."[2]

Protestant missionizing, which had been the dominant form of Christian proselytization in the nineteenth century, took a back seat as Spanish political and cultural control increased. English Primitive Methodist missionaries operated on the island and U.S. Presbyterians were active in Rio Muni. In 1936 the Rio Muni Presbyterian churches were attached to the Cameroon synod. In 1951 Spanish colonial law prohibited further Presbyterian activity but it nevertheless continued. In 1945 Baptists were authorized to open a school for British subjects, especially Nigerians, in Malabo.

Bubi Religion

Officially the Bubi are Catholic, but elements of traditional religion persist in spite of heavy missionizing. In the 1960s a Spaniard observed of Bioko: "The advance of colonization and the progress of the means of transport have modified, in very considerable and, at the same time, substantial form, the modes of living and the primitive forms of settlement . . . but in the highlands of the center and south there are villages in which fresh traces of the past are conserved."[3] The entrance to a traditional Bubi village was described by a Spanish missionary: "Five to ten minutes before arriving at the town, one ran into an arch constructed of unworked stakes, from which hung thousands of amulets, such as sheep tails, skulls and bone of different animals, hen and pheasant feathers, antelope horns, marine and land snail shells, etc., which, as debris of death, revived the memory of their ancestors who lived in the *Borimo* or region of the dead." The arch had at its sides "sacred *iko* trees, with the aim of preventing the entrance into the village of malevolent spirits and preserving themselves from their perverse influences. They also put branches of brackens driven into the ground [at the arch], one of which supports a clay pan of indigenous manufacture; in it they put ever fresh spring water."[4]

The supreme being in Bubi religion is the *morimo*, or the creator of the universe. In the northern part of the island this being is called *Rupé*; in the south he is called *Poto*. Lesser spirits are called *mohs*. The cosmos is divided into three parts. Good spirits live in family groups in the other world and are permitted to visit certain locales on earth. Belief in the efficacy of *mohs* continued through the decades of Spanish rule. Priests (*abba*) were the officiants of the religion. The chief priest dwelled in the highlands of Moka from whence he consecrated lesser priests. The last high priest, or *mote*, died in 1909 and the sacred fire that he maintained was extinguished. Later, even under the heavy-handed cultural imperialism of Franco's Spain, traditional religious practitioners were active; some became involved in the politics of the Unión Bubi.

Religion in Rio Muni

Religion and culture in Rio Muni has been subject to several countervailing influences. On the one hand Rio Muni has absorbed beliefs from a number of other transethnic African religions. On the other, it has been subjected to suppression and subversion by Spanish missionaries, especially during the Franco period. Traditional music and minstrelsy continue, although without the prominence they were given in the immediate postindependence period. The traveling singers who play the *mvet* supposedly have esoteric writings that they show only to participants in their craft. The term *mvet* is also used to refer to the Fang epics. The order of Bebom-Mvet, associated with warfare, had the responsibility of preserving group sagas. The one large-scale epic is the "Cycle of the Chronicles of Engong," in which the immortal people of Engong opposed their enemies, the mortal Okin.[5]

African religion had a resurgence in the period after 1968. In the 1970s religion emanating from Rio Muni played an extremely important part in Equatorial Guinean politics. Since 1979 the role of indigenous religion in political culture has been downplayed. However, it is difficult to imagine that the powerful attraction of African religion has been erased simply through a change of regime.

Biéri, the traditional belief system of the Fang, places heavy emphasis on closeness to the ancestors. It involves necrophagia—parts of a cadaver would be eaten in order to acquire the qualities of the deceased. The skulls of ancestors were displayed by kinship groups; this practice led missionaries to believe the Fang were cannibalistic.

Quadrupeds can be dwelling places for the spirit of a man after death. Thus, antelopes are the dwellings of those who die young, elephants are the dwellings of those who die old. The spirits of women go to dwell in goats. Women are not permitted to eat goats; depending on their age, men are prohibited from eating antelope or elephant.

Religion has been, at times, a manifestation of anticolonial sentiment. It is also an attempt to redress the disequilibrium caused by population decline and tensions between men and women. Especially important are new religions that emphasize their transethnic character. In some instances, changes in religion have involved reinterpretation rather than complete replacement of elements.

The Alar Ayong movement, already mentioned in connection with politics, flourished from the 1930s to the 1950s. It was important in northern Rio Muni and southern Cameroon. By reevaluating the past, especially the mythical past, Alar Ayong sought to create a common charter for the reunification of the Pahouin, including the Fang. Religious leaders, including Christian pastors and catechists, played an important part in elaborating the ideas of the movement. The folkloric and historical interest of the Fang in their past was abetted by the American Presbyterian mission, which published clan genealogies. For instance, they printed the most important mythopoeic overview of the Fang past, *Dulu Bon be Afri Kara* ("The Journey of the Children of Afri Kara").[6]

The *Dulu Bon be Afri Kara* tale recounts the Great Flood and the origin of the African. It also explains the origin of polygyny, woman's relation to man, and the custom of circumcision. Its centerpiece is the strife that led to divisions among the sons of Afri Kara; these sons, in their turn, become the founders of the various groups of Pahouin. According to the account, the Fang migrated from the northeast to an *adzap* tree. Upon reaching it, the Fang chopped a hole in the tree and one by one each member of the migratory band walked through the hole. They immediately found themselves in tropical rain forest. The indigenous pymies aided the migrants. Later the Pahouin groups established their own villages and began to quarrel among themselves. This disunity continued and amplified. Alar Ayong peaked in the 1950s and many of its aims were either subsumed or subverted by African nationalism. The Pahouin remained divided between Cameroon and Gabon in spite of affirmations of past unity.

Another movement, the more strikingly religious Bwiti cult, remains important. In some ways the movements were complementary. In "[Alor Ayong] we see an affirmation of organization and the other, Bwiti, the affirmation of spiritual communality."[7] Bwiti, a transethnic African religion, entered Rio Muni from the south. The new religion complements elements of traditional religion while emphasizing its own transethnic character. Bwiti appears to have originated in the Haute-Ngouiné region of Gabon among the Metsogo and the Masango. In Gabon during the period between 1920 and 1930 the cult increased the number of its "churches" (*mbaza mbwiti*) in the Lambaréné and Kange districts and reached Libreville.

Fang members adopted a corpus of words from the various groups that participated in the Bwiti religion. This is now part of the esoterica that distinguishes members from nonmembers. Bwiti has fractured into many subgroups. Among them are Disumba, Modern Disumba, Asumege Nenig (Commencement of Life), Ndea (Andeanari Sanga), Misema, Ebawga Nganga or Eboga Nzambe, Yembawe, and Mekomb-Kombe.

French colonial hostility toward Bwiti in Gabon was backed up by European missionaries. In Rio Muni the Franco regime in particular treated the religion with special harshness. Bwiti was proscribed and its adherents received harsh prison sentences. In both Rio Muni and Gabon believers moved to fairly inaccessible areas. Until after World War II it was impossible for Bwitist temples to be built near motor roads in Gabon. Persecution led to some division between the public and private faces of the faith. Publicly, under colonialism, it was necessary for Bwiti to avoid offending the European government and religion. A Spanish anthropologist noted that "in our territories, where Bwiti doesn't have a legal existence, the temples are . . . improvised in a clearing in the forest. Sometimes they appear like [other] structures, so that no one may suspect their real purpose."[8] Sermons are preached at midnight and afterward the members commune with the ancestors whose spirits have been attracted to the chapel. During services, which continue until daybreak, the division between the living and the dead is removed. Services reaffirm that the group and its past have not been disturbed.

Bwiti signifies the chief god who reveals himself to the initiated as well as the carved pillar that is the central feature of the place of worship. In areas where the religion is allowed to operate out in the open, the place of worship is usually at the end of the village's central passageway. If the group meets in secret, it is located in a section where members reside. Bwiti temples are constructed over the skull and bones of a noted individual. At the opening of a new temple a sacrifice is offered to the founder, who is recognized as a priest (*nkobe-bwiti*). The essential carved pillar is set up over the skeletal remains in the temple. Members of Bwiti are frequently referred to as "those who have drunk iboga." This phrase refers to a potent drink prepared from the grated bark of a plant (*Tabernanthe Iboga*), which produces hallucinations and sexual desire. The member (*banzie*) drinks the preparation for the first time at his initiation ceremony. Iboga provides the taker with images that cause revelations, which are more important than formal religious instruction. Entrance into the religion depends upon the quality of the image and depth of feeling it evokes.

Dance plays an integral part in Bwiti and other indigenous religions. The religion's major branches have over one hundred fifty songs and more than a dozen basic dances. Bwiti members wear a special costume: "Dances sung by the priest and certain members of the congregation are accompanied by rattles or sistrums . . . for these dances the initiates have to wear a garment made from the skins of tiger-cats and monkeys over a raffia loin-cloth (a reminder of the costume worn by the old Fang warriors), and a feathered headdress, with painted designs on the face and body."[9]

Bwiti beliefs differ from one locality to another and from one ethnic group to another. In Rio Muni the beliefs are found not only among the Fang but also among the coastal Bujeba and Kombe. The religion's theology involves essentially three kinds of stories: those dealing with the origin of iboga, those dealing with cosmogony, and those reinterpreting Christian legends such as the Great Flood or Virgin Birth. In the creation myth Mebege, the creator god, creates the world through a spider (dibobia). Mebege, who himself has predecessors, created the earth and all living things. The origin of humankind comes about through a divine trinity: Zame ye Mebege (god), Nyingwan Mebege (the sister of god and the female principal) and Nlona Mebege (the brother of god). The three create Adam, which is interpreted in folk etymology as a derivation of *adang* (to surmount).[10]

Bwiti differentiates between *Zame asi* (god below) and *Zame oyo* (god above); European attempts to assimilate these ideas into the Christian belief of a good God and an evil Satan were largely unsuccessful. Zame asi was the divinity who presided over worldly success and prosperity. He was also the martial and victorious protagonist of many traditional tales. Zame oyo was more removed and was the "God of spiritual abundance but material poverty."

Among the Ndowe, Bwiti cosmogony is somewhat different and shows heavy Christian borrowings. The god who is the equivalent of Zame asked his sister to help create the human race. Subsequently, he turned against his father, Mwanga (the Fang Mebege), and Mwanga was let down from heaven. Christ was his first son, but played little role in the affairs of the world. Perhaps as a reflection of further Christian influence, in coastal Rio Muni Bwiti also has its Fallen Angel and a tale of temptation; in a war between good and evil, the latter falls into a hole in the earth.

In the face of poverty and infertility, humankind is not powerless. One function of Bwiti is the removal of evil. Population decline and poverty are explained in terms of witchcraft. Society is afflicted with evil. The destruction of old means of witchcraft control makes new means of protection very important. In some Bwiti chapels the traditional Fang evil spirit, *evus*, has been grafted on to Christian elements. Evil is everywhere and leads to antisocial behavior: adultery, drinking, laziness, and other delicts. Evil is fought through demands for correct behavior and ritual practices.[11]

Bwiti emphasizes its ability to counteract the disruptions caused by European contact. One of its aims is to promote *nlem mvore*, "one heartness," among its followers. Unlike some other movements, prayers in Bwiti chapels are offered for the protection of the entire membership not just for members of particular clans. Individuals do recite their own genealogies in their private devotions. But because the genealogies usually end with the name of god Himself, the worshipper is tied to the creator by a consanguineous chain.

The spread of Bwiti among the Fang is considered a reaction to the political and social challenges presented by the seeming superiority of European culture. The movement explains the reason for the period of European domination and says that the Fang are superior to Europeans in spiritual matters. The *Ntagan ye Nsutmot* (Whiteman-Blackman) stories are especially important in this respect. According to one variant, the god active in the affairs of men, Zame ye Mebege, son of the primordial god, Mebege, created the world and left it to go its own way. Later he noticed dissension and returned to impart knowledge to humankind. He first spoke to Whiteman, who listened patiently and received wisdom.

Blackman was impatient and quarreled with his siblings. God was called away before he had finished and never again directly entered worldly affairs. Without the knowledge given by Zame, Blackman was at a disadvantage, which helps explain the colonial episode.

The Fang migrations coastward are frequently reinterpreted in terms of a quest for knowledge. Increased knowledge was gained on the coast, where the Fang came into contact with Europeans. The migration was a transitional period. Before this time the Fang had lived in peace and harmony. According to myth, they had been expelled by a monster crocodile and forced on a long march. When they reached the end of their migration, the coast, there would be a return to harmony and concord among the many Fang clans. Indeed, African spirituality runs counter to European lack of moral fiber and materialism.

Elements of missionary religion are incorporated into some Bwiti temples but are not central. They do not dilute the essential reverence for ancestors. Nonetheless, a number of tales around the central core of the religion reflect missionary teaching. For instance, in one account Adam and Eve committed the sin of sexual intercourse and were expelled from an idyllic forest. Subsequently, Eve gave birth to the ancestors. According to one line of the myth, Adam was later called from the dead and sent to earth in the person of Jesus to teach. One trait of the religion is its tendency to see repetitions and transmogrification of relationships between essential religious archetypes. Jesus is "also and at once the son, the brother and the husband of Mary who is the Old Abel, the Old Eve and the Old and ever-present Nyingwan Mebege, the creative matrix of the world and the primordial source of wisdom."[12]

During his tenure in office, Macias Nguema favored Bwiti. The president was purportedly the son of a *banzie* and during his dictatorship, he managed to acquire considerable magicoreligious power through participation in Bwiti rites, including the taking of iboga. Bwiti was an important counterassimilative ideology and fit well within the anti-intellectual and antimaterialist aims of Nguemist thought. Bwiti, in general, "is contrasted with the European religions which use alien ritual elements such as bread, wine, and candles. This contraacculturative point is frequently extended to a critique of all the material aspects of European life which, it is argued, constitute a threat to the morality and dignity of the Fang way of life."[13]

The Cult of Macias Nguema

In 1977 a human rights investigator for the International University Exchange Fund managed a brief visit to Equatorial Guinea. He reported that "shortly after he became President, [Macias Nguema] revived the

Biéri cult . . . and collected powerful skulls from all over the country."
The president was "presumed to have created sanctuaries for these skulls
in his village at Mongomo." Equatorial Guinea's first dictator supposedly
"collected all the sorcerers and Mvet singers he could get hold and
learnt their Malán (magic)."[14] Macias Nguema took the panther, called
"el tigre" by the Spaniards, as his protective animal and symbol. In
some quarters it was believed that he could transform himself into this
animal in order to revenge himself against his enemies. The investigator
found that the president made extensive use of Mvet singers and used
them to disseminate his propaganda. New dances, some supposedly
choreographed by the president, were introduced to Bwiti. It is significant
that at his execution, Moroccan troops had to perform the task because
Equatoguinean troops were still intimidated by the magical powers
surrounding Macias Nguema.

Traditional religion was an obvious ideological antidote to the
Hispanicization of the Franco period. Shortly before independence,
Monsignor Rafael Nzue Abuy, a Fang clergyman, declared: "We . . .
warn you against anti-religious rabble-rousers. He who does not fear
God will not respect the laws of Human Rights."[15] The Catholic clergy
were considered important foci of opposition and were subjected to
increasing repression after independence. In 1972 Nzue Abuy was exiled
and government officials began emphasizing the slogan, *"No hay mas
Dios que Macias"* ("There is no other God than Macias"). A party chant,
"God created Equatorial Guinea thanks to Macias—Without Macias
Equatorial Guinea would not exist," became obligatory in church services.

In 1973 the ruling party said that "the Catholic Church, during its
presence in the Republic of Equatorial Guinea, has always been a faithful
instrument at the service of colonialism, plotting schemes which apparently were considered religious, and had knowledge of the constant
campaign developed clandestinely against the President for Life of the
Republic."[16] Many church buildings were destroyed, particularly those
in the interior of Rio Muni. The homes of church members were raided
and baptismal certificates, religious images, and birth certificates were
seized. The cathedral in Malabo and church buildings in other locales
were taken over by the state. In Bata the government appropriated a
museum, cars, and houses belonging to the church.

In 1973 Macias Nguema also de-Europeanized most of the place
names in the country, thus depriving them of both colonial and Catholic
connotations (e.g., Santa Isabel became Malabo). The following year a
decree ordered priests and pastors to open their sermons with the words
"Nothing without Macias, everything for Macias, down with colonialism
and the ambitious."[17] In Bata in 1975 the head of state informed an
audience that "false priests" were "thieves, swindlers, exploiters and

colonialists."[18] The populace was warned that contact with Catholic clergy would be severely punished. Catholic schools were effectively shut down by a decree banning all private education. The head of state was proclaimed the "unique miracle" of Equatorial Guinea. Before 1978 church services were made to include sycophantic adulation of the president-for-life. Political rhetoric evoked the image of the leader as a miraculous figure, eclipsing discredited Roman Catholic saints. In May 1978 the last Spanish missionaries were expelled and the country was proclaimed an "atheist state."

Religion During the Present Regime

At present Bwiti is under a cloud in Equatorial Guinea and under attack in Gabon. One branch, the Equatorial Guinea–based *Mvoe Ening* ("the quiet life") movement, is proscribed by President Omar Bongo. This followed admission of anthropophagy and murder on the part of religious officiants in 1988.

After 1979 the new regime of Obiang Nguema appeared to distance itself from indigenous religion and anticlericalism. Visitors seldom hear open talk about Bwiti. However, it is doubtful if certain groups, for instance the Esangui of Mongomo, have completely abandoned the cult that affirms their collective identity. Traditional religion continues in Equatorial Guinea beneath a cover of silence and a professed return to the church fostered in colonial days. In 1987 Pedro Nsue Ela, sometime director-general of consular and cultural affairs, acknowledged:

> There exist Fang, Bubi, and Ndowe customs that are attributable neither to one concrete individual nor to a determined epoch. The practice of witchcraft, the *morimó* and the *butí* [sic] continue spreading everywhere as before, now and afterwards, in all the national territory.
>
> The Fang, of whatever tribe or district, in their matrimonial rites continue charging bride-price, which includes at times the sacrifice of a human being.
>
> The Bubi and the Ndowe welcome *morimó* and *mocucu* in their catacombs in order to adore the dead and purify their spirits.
>
> The rulers continue to welcome the magical cures of the wizard and herbalist in order to obtain public posts and to alleviate their worries.
>
> Thus is our society. Thus is the reality of the Guinea that we have inherited from our ancestors. The Church wanted to combat these practices and was not able.[19]

Obiang Nguema, who received part of his education in Spain, has portrayed himself as a faithful son of the church. According to a U.N. investigation shortly after the fall of his predecessor, "the Church is not only permitted to maintain educational institutions but the Government itself has called upon priests to give classes in religion in its own institutions."[20] In February 1982 Pope John Paul II visited Malabo and Bata, partially in recognition of Equatorial Guinea's place in Catholic Africa. At present, it was estimated that 86 percent of the Guinean population is Roman Catholic.[21] The present orientation of the government has led to a reinterpretation of aspects of traditional Fang religion. For instance, the highest African cleric in Equatorial Guinea, Rev. Rafael Nzue Abuy, sees traditional morality and theology as essentially congruent with Catholicism.

Non-Catholic denominations continue. At present the Methodists have reoccupied their old mission church in central Malabo. Toleration is not complete; in early 1986 Jehovah's Witness meeting halls were abruptly closed.

EDUCATION

Because of the intensive Hispanicization of the Franco period, culture and culture conflict have played a major role in Equatorial Guinea. European education sought, as elsewhere, to inculcate the values of the metropole. Spanish colonialism stamped the colony at least superficially with its culture. By the late 1920s, 903 boys and 282 girls were receiving secondary education through the Roman Catholic missions. Five thousand boys and 1,034 girls were receiving primary education. In order to create loyalty to the government, African youths were taken to Spain for education with the avowed intent that, on their return, they would be colonial propagandists. Within the colony, the emphasis was on vocational education.

In 1914 an *escuela external* (secondary school for nonboarding pupils) was opened. Between 1914 and the late 1920s, 609 students were matriculated. Of these, 26 were European and 210 were Bubi; there were 113 Fernandinos and 75 students from Cameroon. There was also a scattering of boys from Rio Muni, Lagos, Calabar, Monrovia, Sierra Leone, São Tomé, and other places. Among the school's most notable graduates was the Bubi Apolonio Eria, who obtained a teaching certificate in the metropole and went on to serve as an auxiliary teacher in the Official Children's School in the colonial capital.

In 1935 a teacher-training institute was established in Malabo. It was underfunded and the auxiliary teachers it produced received only six months of training. Although in theory schooling was compulsory for ages five to fifteen, in 1941 only 111 assistant teachers provided education. A true educational program, based on metropolitan paradigms, began in 1942. A good secondary school, the Instituto Cardenal Cisneros opened in the capital. In the following year an educational system was established by law. Heriberto Alvarez Garcia was inspector-general of education and head of the Escuela Colonial Indigena from 1944 to 1955. The training of auxiliary teachers and administrators was raised to the baccalaureate level. An attempt was made to raise the salaries of African educators, but Madrid opposed the idea. From 13,900 primary school students in 1950–1951, the number increased to 48,000 in the late 1960s. In 1958 a technical school was opened in Bata under the supervision of the La Salle Brothers missionaries. Five years later, a normal school was finally established in the capital of Rio Muni.

By the late colonial period Spanish Guinea had one of the highest ratios of schools to population in sub-Saharan Africa. In 1966 there were 147 elementary schools, with 21,421 pupils and 32 upper-level primary schools with 1,565 pupils. The two secondary schools, one in Bata and one in Malabo, had 986 students. The educational record of Spain in these years, especially after 1959, should be looked at in context. Results may have been less dramatic than they seemed: very few students continued beyond primary school and most of the auxiliary teachers were undertrained. A disproportionate number of the students counted in statistics were Europeans and few students did well on academic tests.

During the first postindependence government, education was not emphasized. The regime cast about for a viable educational philosophy and came up with a "third way" between the systems of the Eastern and Western blocs. In 1970 the education ministry outlined a program that rejected Western "liberal education." It also refused to accept an educational system molded by Marxist ideology. There were three ways of approaching the issue: the individualist-idealist position, the collectivist-materialist position, and "the position that consists of considering the individual, but not individualism, and in considering society, but not collectivism." The "individualist-idealist" position was opposed as a product of the sterile order that had given rise to European colonialism. The collectivist-materialist position was rejected as "the position of the Jew Carlos [Karl] Marx, according to which the individual ought to integrate himself completely within a total collectivity." The ministry noted that "in reality, this theory has never been applied in any country in its extreme conception; and within the purely Communist countries the way is opening [for] the idea that the individual counts more and more."[22]

In 1972, 360 primary schools served 35,902 students and had 578 teachers. Spain contributed to the construction of new secondary schools in Rio Muni, but only two, in Mongomo and Ebebiyin, functioned. Education took on an increasingly Maoist vocabulary. In 1973 the Ministry of Education was renamed the Ministry of Popular Education, Traditional Arts, and Culture. Teaching declined greatly; much schoolroom activity centered on the repetition of the principles of the "great leader," "mass physical training," and revolutionary songs. Private schools were closed. Cuba sent in personnel upon invitation. The Cubans taught in the Escuela Normal (the teachers' training college) and in the primary and secondary schools. However, because they did not work out well, they were withdrawn by their government.

Anti-intellectualism was the cornerstone of Macias Nguema's cultural policy. He fired approximately 600 teachers, which corresponds to approximately 600 single-teacher schools. In 1972 anti-Macias leaflets were circulated at the Instituto Carlos Lwanga in Bata. In the aftermath of these leaflets, some teachers, students, and parents were arrested. Minister of Education Eñeso Ñeñe was publicly executed and a number of educators thought to be disaffected were killed. From 1969 to 1976, 75 educators and educational administrators were killed. This included three ministers, a secretary-general, and a director-general of education. The use of the term *intellectual* was prohibited and in 1973 a member of the cabinet was fined for using it.

Today formal education is beyond the hopes of most Guineanos. In 1988 it was estimated that only 55 percent of the population was literate.[23] According to another study, illiteracy affects a far greater percentage of the population.[24] School buildings are in a state of disrepair and there is a severe shortage of trained personnel. In accordance with priorities established by Equatorial Guinea, the UNDP established a project for the training of teachers and improvement of the curriculum. The project was carried out in the period 1983–1987. It was evaluated in 1986 and the United Nations concluded that "the project operated in very difficult circumstances."[25] For example, sometimes classes contained 50 to 160 students with marked differences in ages. There was also a lack of teaching materials and a very low level of teacher preparation. In the same year primary and secondary education covered a geographical area of 28,000 square kilometers composed of 18 districts, divided into 695 primary education centers and 9 institutions of secondary instruction. One thousand sixty-three primary school teachers provided education for some 58,380 pupils; a national average of 55 students per teacher. One hundred ninety-five teachers provided instruction for 5,287 secondary students; the student/teacher ratio at this level was 27 students per teacher. Secondary education was found only in 9 districts.

In spite of budgetary constraints, Equatorial Guinea has tried to increase the number of students. In the school year 1984–1985 to 1985–1986, 165 posts were created in primary schools; this was an 18 percent increase. One serious problem is the lack of secondary education and

its poor interrelation with primary schooling. In addition, it is difficult
to determine the number of years that constitute primary versus secondary
education.

The UNDP has attained tangible results in qualifying teachers and
in training pedagogical counselors. A curriculum project for grades one
through five has been completed. The program had major cofinancing
from the World Bank (the International Development Agency); the total
allotted for the project was $6.4 million. The U.N. educational effort
continues through the period 1987–1991. Emphasis is on restructuring
and strengthening the National Center for Teacher Training. A priority
is the training of teachers, school inspectors, and others connected with
school administration. The curriculum drawn up for the first three years
of primary school will be disseminated. A preschool curriculum is being
formulated and preschool teachers and monitors are being trained. The
U.N. project plans a contribution of $940,000.[26]

The training of public servants will continue. In 1981 the Martin
Luther King School of Administration opened in Malabo. The World
Bank and the UNDP are collaborating on a project aimed at improving
the training and activities of the Escuela Nacional de Administración
Publica (National School of Public Administration) so that the school
will be able to give courses and seminars to upper-level administrators.
The school will also be empowered to develop consultant activities vis-
à-vis other national institutions. The project is designed to give the
school the capacity to undertake basic research.[27]

Other agencies also have ongoing educational projects. Spanish
Technical Cooperation maintains activities on a large scale, especially
through the provision of expatriate professors. In 1981 the Miramar
Hotel School was opened with Spanish assistance. It offers accelerated
courses for adults and trains electricians, typists, and plumbers, among
others. France gives support to primary education, principally through
an allotment of teaching materials and other equipment. As for the
physical infrastructure of the educational system, the World Bank and
African Development Bank are aiding in the purchase of materials for
the rehabilitation and construction of primary schools.

LANGUAGE

Today the majority of Equatorial Guinea's African ties are with its
Francophone neighbors. Their economic pull has led to speculation that
the country might change its official language. The presence of dozens
of French technical assistants in Malabo makes this plausible. There is
a French cultural center and, no doubt, France hopes to extend the
range of its operations in the future. In September 1988, during Obiang

Centro Cultural Hispano-Guineano, Bioko. (Photo courtesy Charles W. Grover)

Nguema's visit to Paris, it was said that as "the only Spanish-speaking Bantu country in equatorial Africa . . . we feel like orphans [because] the other Bantu countries of this region all are French-speaking."[28] The president added that Equatorial Guinea would seek entrance in "francophonie." Paris was pleased to note that French was a required language in the schools and, in many spheres, it was becoming the working language.

Spanish remains engrained among the governmental elite. In 1982 a Spanish-Guinean Cultural Center (Centro Cultural Hispano-Guineano) opened. It contains two libraries, one Spanish and one Guinean. Art classes are taught in the center and there is gallery space for the display of paintings and sculpture. The center sponsors sports teams and has classes in photography and music. In addition, its auditorium is the center of cultural activity. In 1989, in an effort to further bolster its cultural position, Spain agreed to construct a new secondary school in the capital.

In 1984 the first anthology of Equatorial Guinean literature was published with the collaboration of a Spanish publishing house.[29] In the same year the first International Hispanic-African Congress took place in Bata. The congress debated the role of European culture in African states with special reference to Equatorial Guinea; results were mixed. A second congress, scheduled for Spain in 1985, did not take place.

However, notice was taken by the Fourth Conference of Ministers of the ten member states of the Centre International de la Civilisation Bantu (CICIBA). The ministers promised future cultural collaboration with the one non-Francophone state in the region.

In the mid-1980s, *Africa 2000*, a cultural and historical magazine, began publication. It is sponsored by the Centro Cultural Hispano-Guineano and is of consistently high quality. The centro has also published several literary works, among them Juan Boneke Balboa's *Sueños en mi selva, antología poética* (Dreams in my forest, a poetic anthology) and Manuel Fernández Magaz's *Cuentos en el Abáa* (Stories in the Palaver house); both were published in 1987.

Some Guineanos defend the historic mission of Spain and, by extension, Equatorial Guinea. In 1987 Anacleto Olo Mibuy, writing in the cultural center's journal, opined that Africa had room for an *"espacio hispánico o Afroibérico."* This cultural space would allow Africa to benefit from the merits of Spanish civilization: "The contribution of the Iberian Peninsula to the knowledge of new worlds is . . . unquestionable . . . at the same time . . . it demands the assumption of responsibilities and of a certain pride at having been, accidentally, united in Africanness and *hispanidad*.[30] Somewhat more defensively, Donato Ndongo-Bidyogo, the editor of the center's publication, says that Hispanophilia should not be "an object of speculation, because, among other reasons, in our geographical and ethnic jumble, few Cameroonians question their belonging to the linguistic and geopolitical area of France; and every year we see the Nigerian leaders in the photographs of the Commonwealth . . . and no one is shocked at Lagos."[31]

THE STATUS OF WOMEN

In 1986, in a report on human rights, the U.S. government concluded that "many times more males than females enter secondary school" and a "higher proportion of the graduates is male."[32] In 1988 a women's organization noted that:

> She produces the 90% of the food used in this Country and sometimes part of what she gets from her farm must be sold in order to buy essential commodities. She starts this type of activities [sic] at the age of 7 or 10 and then when she is 40 or 50 years old she would not be ablle [sic] to work any more because she has been overworked all her life. . . . The worst of all, after her death, [a] collection has to be made to buy her coffin. Her life is tragic many a time [because] she has to go to work on the farm ill with malaria.[33]

In Equatorial Guinean society women continue to work in agricultural and other nonhousehold pursuits. On the mainland, one writer has said that "compared with many other Bantu tribes, the Fang appear to have paid little attention to farming." In addition, "subsistence agriculture was, on the whole for the women, while the men devoted themselves to trading, at time raising the necessary capital through wage labour in the timber industry, and to the sporting events of hunting and . . . raiding."[34] This sexual division of labor is not peculiar to Equatorial Guinea. What particularly exacerbates tensions between the sexes is declining population and its impact on patterns of work and marriage. Greater mobility of both men and women has also led to a rise in prostitution.

By the 1940s many of the old ways of life were being gradually impinged upon. The entrance of new goods began to modify aspects of traditional culture, including dress. In the precolonial era, clothing was a much clearer delineator of status. The favorite or wife of a chief wore an elaborate hairdo that distinguished her from the other women of the village. Also, an important woman's arms and legs were adorned with metal bracelets and anklets. Heavy collars were worn around the neck. Adornment was completed by a chain of little bones and shells that ran from ear to ear passing through the cartilage of the nose. When working most women wore a cylindrical basket made of bamboo or a similar material supported on their backs by a strap.

The past economic and cultural context does greatly influence the present condition of women. Many women did not have children; infertility caused some men to accuse their spouses of promiscuity. New economic conditions also upset traditional sex roles. As trade increased, "most of the traditional skills at handicrafts disappeared and the capitalisation of trade goods upset the circulation of marriage payments which, in turn, weakened the social organisation." This economic change "is relevant to the present hardships in Equatorial Guinea in three ways: Non-existence of the cottage industries which might have provided basic necessities, adherence to cash economy in spite of runaway inflation, and serious social disorganisation."[35]

It is interesting that Bwiti attempts to ease increasing conflicts between men and women. Harmony "is accomplished by a ritual which is symbolic of sexual activity. . . . In this symbolic production and celebration of creativity, the women play an important role. They perform such rituals as introducing the female principle of creation by bringing in the substance of the female deity of the universe, Nyingwan Mebege, from a sacred pool in the deep forest." Through the mediation of religion, "men and women together ritually create a universe, a microcosm as it were, in which they can dwell with dignity, with a sense of moral adequacy, and in one-heartedness."[36]

Unfortunately, such ideological adjustments do not change the woman's socioeconomic role. During the 1970s restructuring was promised and a women's wing of the only political party was established. In a meeting of the civil government of Rio Muni in 1975 the Women's Revolutionary Section of the party was told that the Central Committee had "abolished totally the ill-treatment of women except in the cases of subversion." Furthermore, it was stated that "for work in the fields the Great Master and Father of our Revolution recommends that women wear trousers to allow free movement and to protect the body." In phrases that were to be echoed later, it was announced that "the Liberator of our Great People . . . in his never-ceasing determination to place the woman at the summit of progress, once more invites her to prepare herself politically in the way he has outlined and considers invalid any meeting or association in which she has not participated."[37]

In spite of the president's statement, few women rose to positions of leadership. Macias Nguema's first wife, Clara Osa, became director of pharmacies before fleeing to Gabon in 1973. Some women, for instance the wives of Ondo Edu and Buendy Ndongo, shared the fate of their spouses and were liquidated during the terror. As a result of complaints about forced labor, in August 1976 the U.N. Commission on Human Rights studied the condition of women. Investigation showed that women were consigned to hard physical labor, especially if they refused the advances of government officials. The Alianza Nacional de Restauración Democrática created a women's organization for monitoring the situation.

In 1981, President Obiang Nguema stated his desire to give women a greater role in national life: "The advancement of women . . . signifies the firm intention of women to break with the past 'social taint' of relegation and semislavery, and her stepping forward in society as a being . . . capable of the highest tasks and difficult responsibilities; a being capable of meriting, for her value and capability, to march on a par with man."[38] In the same year Obiang Nguema named Frieda Krohnert, an ex-wife of Macias Nguema, to be director of national health facilities. In spite of such appointments and pronouncements, women continue to be politically and economically subordinate.

In 1985, partially in an effort to encourage an influx of foreign labor, Equatorial Guinea ratified labor legislation that promised equal pay for equal work and maternity benefits. At present the most important positions held by women include a mayorship (Luba) and the positions of vice-minister of labor and vice-minister of health. Malgalena Penda Recu is a prominent member of the national legislature. The country has an Association of Equatorial Guinean Women. In addition, the

Dirección para la Promocion de la Mujer (Office for the Advancement of Women) is to date the most tangible evidence of new government commitment. The director, Purificación Angué Ondó, is the regime's most outspoken member on women's affairs. Angué Ondó noted that women "are in the Assembly, in the Government, in the Embassies, in the city halls, in the courts, in the town councils, deliberating and speaking at the side of their brothers, in order to find solutions to the problems that weigh down our Country." In 1987 she strongly attacked the writer Agustin Nse Nfumu for "affirming the existence of prostitution in Equatorial Guinea and presenting it as the only way of life for the Equatorial Guinean woman."[39]

The president's wife, Constancia Mangue de Obiang, has a handicrafts center for women in Malabo. In March 1988 Hilda Adelfarasin, president of the Nigerian National Council of Women's Societies, visited Malabo and formed a joint Nigerian–Equatorial Guinean women's committee. In the same year Trinidad Morgades Besari was named secretary-general of the republic's Council of Scientific and Technological Research. She is also one of the founders of the Christian Women's Association.

Several private aid organizations have programs that directly or indirectly benefit women. During the period 1984–1986 the U.N. Development Program carried out a small enterprise program in cooperation with the International Labor Organization. Participants were trained in cloth dying, sewing, and the making of bamboo furniture. At the same time eighteen delegates of the Office for the Advancement of Women were trained in planning techniques and programming.

The UNDP has proposed that the present three-year plan be continued and that support be lent directly to female agricultural groups and small-scale artisans.[40] The development program is also working with the Office for the Advancement of Women in an effort to improve its efficiency and capacity to initiate programs. The three-year project has a contribution from the UNDP of $400,000. The project of women's enterprises, specifically, has tight coordination with the Livestock and Garden Produce Project in the district of Bata (European Development Fund) and with the Integrated Rural Development Project of Niefang (bilateral French cooperation).

A project most beneficial to women is the development of rural cooperatives. Men and women belong to separate cooperatives. Women's cooperatives concentrate on the production of food crops: plantain, malanga, bush pepper, tomatoes, and pineapples. There is some promise that in the future the country will be able to export food crops to its neighbors such as Cameroon and Gabon. In that event, the purchasing power and well-being of Equatoguinean women would be greatly enhanced and could herald new departures for the economic future of the country.

HEALTH

A low birthrate is a continuing problem in Equatorial Guinea. As seen, by the 1940s, Bioko shared most of the features common to tropical island economies, save one:

1. casual seasonal employment for a large part of the labor force;
2. limited investment and employment opportunities outside the export sector; hence a lopsided economy;
3. a situation of economic dependency. The most obvious feature of this dependence is the extent to which such countries depend on exports and imports and the very small control that they have on the terms under which this trade is carried out;
4. a preindustrial social system characterized by relatively impermeable divisions, an introverted mentality on the part of the different groups and low social mobility;
5. great inequalities in the distribution of income and wealth;
6. galloping demography on a relatively static economic base; hence a growing rate of discrepancy between the rate of population increase and the rate of economic growth.[41]

Bioko was deviant in the last respect. The indigenous population declined until World War II and then began a slow increase. In the 1940s a Spaniard noticed "a new period of [agricultural] expansion with improvement in yields, in which the Bubi population now participates, compensating for the decadence of the Fernandinos."[42] This turn of events would have seemed almost impossible earlier. According to one set of statistics, in 1936 the indigenes numbered 9,352, or 36 percent of a total Bioko population of 25,770. This number was maintained, with almost no variation, in the 1942 census (9,350 Bubi); the total population rose to 33,980, of whom 30,661 were black. In 1942 indigenes were 30.4 percent of the African population and 23.5 percent of the total population. Eight years later the Bubi population had grown to 11,355. This was only 28.9 percent of the total population.[43]

A 1942 study said that, out of 333 pregnancies on the island, "there were 38 abortions (11.3 percent) and 43 still births (12.0 percent), 85 children died before the second year (25.5 percent) and 28 between the second and the fourteenth (8.4 percent); whereby there were 194 children lost (58 percent) against only 139 alive (42 percent)." This situation presented a severe problem for future development. Spanish medical personnel who conducted a health survey published in 1942 commented that many diseases had been conquered: "A great many causes . . . that in the past prevented the increase of population— endemic diseases: epidemic small-pox, trypanosomiasis, malaria, yaws, etc.; . . . slavery, lack of protection to workers—have disappeared or lessened because of colonization." What the health surveyors could not understand was continuing underpopulation. The team noted that "the population has not increased as much as was to be expected, the reason being that simultaneously with these propitious improvements, Europeans have carried some other causes of depopulation—gonorrhea, syphilis— which quickly spread, becoming a new calamity for offspring. These, together with malaria, are the endemic diseases we must try to attack in order to attain an increase in the population of blacks."[44] The postwar period was characterized by the decline of once prevalent diseases and their replacement by new ones. For example, illnesses like trypanosomiasis seemed practically eliminated. Whereas at one time 43 percent of the population had been reported infected, by 1949 the rate was only one case for every 4,000 inhabitants.[45] Deaths due to smallpox and yellow fever were also reduced.

Demographically, Equatorial Guinea continues to differ from many of the island nations of the Caribbean and elsewhere. In 1986–1987 the rate of infant mortality was 144 per 1,000 births and the gross mortality rate was 20 per 1,000 births. Life expectancy at birth is approximately 44 years. In the late twentieth century "the Atlantic fertility had fallen in the Caribbean states by about 10 crude birth rate points from the range 31–45 to 20–35/1000. . . . Birth rates in the low to mid-30s/1000 characterized Gabon and Equatorial Guinea, but both could be attributed to the pathological conditions which affect large populations in this part of Africa."[46]

Demographers have long assumed that the region including Equatorial Guinea and its neighbors has been undergoing certain trends: "It was reported widely that there was a wide belt of relatively low fertility extending from Cameroon to Gabon, North Western Zaire, Central African Republic, Southern Sudan and to Western Uganda. . . . It was also widely assumed that it was a consequence of locally endemic diseases, especially gonorrhoea."[47] Although the provision of health services in the region may have made any such generalization unwise, it seems that it applies very well to the past demographic history of Equatorial Guinea. The lineaments of the demographic situation are evident: "The main features of this demographic crisis . . . are as follows: a decline in population, varying according to district and circumstances; an imbalance between the numbers of men and women—with the sex ratio rising rapidly among the older age groups—and a more rapid aging of the female population; and an inadequate demographic base, due to the high infant mortality."[48]

Infant mortality statistics are somewhat inconsistent. In addition, they are compiled from hospital statistics in a country with inadequate medical care. At the beginning of 1985, infant mortality began to rise after a long period of decline. The rise began in the rural areas and is now seen in urban centers. The gross mortality rate was estimated at 20 percent in 1986. The infant mortality rate has continued to increase since 1983 and in 1986 was estimated to be in the range of 144 per 1,000 births. The maternal mortality rate was an estimated 4 percent. If this data is accurate, Equatorial Guinea has a very high infant mortality rate.

Newborn deaths are also a serious health problem. In the period from 1981 to 1986, 280 infants died out of 10,134 births. This represents an early postpartum mortality rate of 27.63 per 1,000 births. Deaths of overdue fetuses were also prevalent. Of the 10,134 births registered in the Malabo Maternity Hospital in the five years following 1981, 252 overdue fetuses died (22.89 per 1,000 births). In the case of premature infants the record was better. Of 9,902 premature births in the Malabo hospital in the five-year period, only 48 of these infants died (4.8 per 1,000 births).

Several diseases greatly influence the viability of the Equatoguinean infant. Information from the general hospital in Bata in the mid-1980s shows that malaria is especially dangerous between birth and one year of age. After age one, the danger gradually declines. Although the number of cases of malaria topped all others, measles was the greatest cause of infant death. In the Bata hospital, measles caused 9 deaths, whereas malaria resulted in 5 deaths.

Diseases, especially childhood diseases, are seasonal. For instance, the risk of infant death is higher in Rio Muni during the dry season when the population drinks unsafe water. In February, March, April, and part of May, and from October to December, the malarial mortality rate goes up considerably. Measles increases during the months from February to May. In the rainy season cases of infantile respiratory disease increase as do deaths from gastrointestinal ailments. Trypanosomiasis, which once appeared to be on the wane, is reported in Kogo, Ntem (Rio Campo), and Anguma Ncoasas. In rural areas "one of the principal causes of newborn deaths (those less than four weeks old) is attributable to tetanus, low birth weight and to other diseases which are due to cultural, traditional and ethnic factors particular to the rural milieu." No doubt these "cultural" factors might be overcome with a conscientious and effective program of public health education. In the countryside,

Luba Hospital, Bioko. (Photo courtesy Charles W. Grover)

"for children up to the age of 5 years, deaths are attributable to malaria, diarrhea, anemia, respiratory ailments, measles and malnutrition."[49]

On Bioko diseases are most prevalent from February to April. Children are greatly affected by water-borne ailments such as diarrhea, typhoid fever, and hepatic amebiasis. In Luba and its environs there have been outbreaks of trypanosomiasis during the rainy season. The area around Malabo (Campo Yaoundé, Elá Nguema) has a high rate of water-borne diseases and malaria. Overall, Bioko appears less healthy than Rio Muni because of the continued prevalence of drepanocitosis and trypanosomiasis.

A serious problem has been the collapse of the medical infrastructure and the resurgence of certain diseases. Before independence Equatorial Guinea appeared to be a model colony. By the late 1960s there were hospitals with 100 to 345 beds in many towns: Malabo, Bata, Mikomeseng, Luba, Kogo, Evinayong, Ebebiyin, and Mongomo. The colony had a ratio of 1 hospital bed for every 315 inhabitants. The quality of the care doubtlessly varied. In some border areas the population preferred to cross over for medical care. This is still the case; in 1987, for example, the relative healthiness of the people of the Ebebiyin district was attributed to their ability to cross into Cameroon for medical care.

In 1979 the United Nations reported that hospitals, like most other institutions, were in a sad state. For instance, the hospital in Luba was completely empty. Before independence it had been run by nuns and had space for 280 patients. The maternity ward had been particularly good. In late 1979 the hospital had no running water, no electricity, and no medical equipment. A small pharmacy had a few drugs. The hospital in Niefang was in worse condition; part of the structure had caved in and services were limited to outpatient care.

Further problems awaited in the 1980s. Malarial outbreaks arose again as chloroquine-resistant strains of the disease appeared. In the mid-1980s cholera appeared, as it had already done in 1973 on Annobón. AIDS information is not available at the present time, but the government is aware of the danger. In 1987 it created the National Committee of Prevention of AIDS and a technical commission to assist the committee. The national committee is made up of persons from different ministries who understand that the problem, like any other health problem, should be attacked in a multifaceted and multidisciplined way. The national committee's principal function is to formulate a strategy in conjunction with the technical committee. Following guidelines set out by the World Health Organization, the committee's first task will be to find out how widespread the virus is in the republic.

Obviously there has been a decline in living conditions—nutrition, shelter, etc.—and the availability of health care. At the end of the colonial period, a foreign nutritionist remarked: "Equatorial Guinea seems to be an island of modest prosperity in the Middle African sea of malnutrition and poverty. The fact that the literature is poor in documents on malnutrition seems to indicate that at least there is no serious problem of food supply and distribution." In 1963 the colonial government conducted a study of diet and dietary habits. This study found that 72 percent of the calories on Bioko and 79 percent in Rio Muni came from carbohydrates. Seventeen percent on Bioko and 16 percent in Rio Muni came from fats. Proteins provided 11 percent of calories on Bioko and 5 percent of calories in Rio Muni. It was concluded that "compared to other regions of tropical Africa, Fernando Póo [Bioko] comes after Togo and Liberia in the amount of calories provided by carbohydrates, while Rio Muni comes sixth. The total caloric intake in Fernando Póo is relatively high at 2,470 and so is the intake of total protein at 67.6 grams per capita per day (34 grams of animal origin)." On the mainland the situation was not so good; out of a total caloric intake of 2,252, protein intake was only 30.2 grams.[50]

After the traumatic years 1969–1979 the pattern appears to have altered but not drastically. Malanga (Xanthosoma) grows in abundance. The problem in the late 1980s is to provide for its transport to areas of relative need. Food shortages have been reported. In 1982, after his visit, the Pope affirmed that malnutrition existed in Equatorial Guinea.

Later this view was affirmed by the U. N. Food and Agriculture Organization. At the end of the dry season, food becomes scarcer and less varied. Malnutrition is present, particularly among children. Diarrhea becomes marked among those under 5 years old. Thus far the situation does not appear to be uncontrollable. Food scarcity is not so marked in rural areas where malanga and other subsistence crops enable a family to manage. This may change; the rural population is growing older due to the out-migration of the young.

The small number of inhabitants and the relative abundance of indigenous food crops militate against runaway famine. However, one cannot guarantee that political factors will not interfere with the planting and harvesting of the means of subsistence.

Medical assistance is now supported by diverse bilateral cooperation. Donors provide training and medicines as well as medical personnel. The government of Cuba has been involved in providing primary health care. Plans are afoot for the future. Italy is sponsoring a project that will improve the water supplies to Bata. The World Bank is funding garbage disposal and sewage projects in both Bata and the capital. A two-year antimalarial program will cost approximately $500,000. It is hoped that between 1988 and 1992, the number of grave cases and deaths can be reduced by 50 percent. Areas of intervention in basic sanitation and the provision of potable water have been identified by the U.N. Development Fund. Compared to that available in the 1970s, medical, medical care has also improved. Nevertheless, a scarcity of qualified functionaries and deficiencies of know-how continue. Monetary policies have restricted the size of the administrative apparatus and the low salaries serve as a disincentive.

NOTES

1. Dare Babarinsa, "The Rise and the Decline," *Newswatch* (Nigeria), May 23, 1988, p. 16.

2. René Pélissier, "Spain's Discreet Decolonization," *Foreign Affairs* 43, 3 (April 1965), p. 525.

3. Manuel de Teran, *Sintesis geográfica de Fernando Póo* (Madrid: Consejo Superior de Investigaciones Cientificas, Instituto de Estudios Africanos, 1962), p. 44.

4. Antonio Aymemí, *Los bubis en Fernando Póo, Collección de articulos publicados en la revista colonial la "Guinea espanola"* (Madrid: Galo Saez, 1942), p. 131.

5. Max Liniger-Goumaz, *Connaître la Guinée Equatoriale* (Geneva: Editions des Peuples Noirs, 1986), p. 27.

6. James W. Fernandez, "Fang Representations Under Acculturation," in *Africa and the West, Intellectual Responses to European Culture,* ed. Philip Curtin (Madison: University of Wisconsin Press, 1972), p. 40.

7. James W. Fernandez, "The Affirmation of Things Past: Alar Along and Bwiti as Movements of Protest in Central and Northern Gabon," in *Protest and Power in Black Africa,* eds. Robert Rotberg and Ali Mazrui (New York: Oxford University Press, 1970), p. 440.

8. A. de Veciana Vilaldach, *La secta del Bwiti en la Guinea Española* (Madrid: Instituto de Estudios Africanos, 1958), p. 48.

9. Georges Balandier, *The Sociology of Black Africa: Social Dynamics in Central Africa* (London: Andre Deutsch, 1970), p. 226. Also see Carlos González Echegaray, "Aportación al estudio de la canciones y danzas de la provincia española de Rio Muni," in *Estudios Guineos, Vol. 2, Etnologia* (Madrid: Instituto de Estudios Africanos, 1964), p. 143.

10. James W. Fernandez, *Bwiti: An Ethnography of the Religious Imagination in Africa* (Princeton, N.J.: Princeton University Press, 1982), p. 333.

11. Fernandez, "The Affirmation of Things Past," p. 446.

12. Fernandez, *Bwiti,* p. 340.

13. Fernandez, "The Affirmation of Things Past," p. 447.

14. Robert af Klinteberg, *Equatorial Guinea—Macías Country* (Geneva: International University Exchange Fund Field Study, 1978), p. 53.

15. *Ibid.,* p. 51.

16. Partido Unico Nacional de Trabajadores de la República de Guinea Ecuatorial, *Resoluciones generales del tercer congreso nacional* (Bata, 1973), p. 4.

17. Alejandro Artucio, *The Trial of Macias in Equatorial Guinea: The Story of a Dictatorship* (Geneva: International Commission of Jurists and International University Exchange Fund), p. 12.

18. *Ibid.*

19. Pedro Nsue Ela, "Cartas de Nuestros Lectores," *Africa 2000,* año 2, epoca 2, num. 2/3 (1987), p. 48.

20. Fernando Volio-Jiménez, *Study of the Human Rights Situation in Equatorial Guinea,* Unesco, Commission on Human Rights, 36th sess., Item 12, E/CN.4/1371, February 12, 1980, p. 43.

21. U.S. State Department, *Background Notes, Equatorial Guinea* (Washington, D.C.: U.S. Government Printing Office, 1986), p. 1.

22. Ponciano Nvo Mbomio, "Relaciones entre la educación y el desarrollo económico," *Organo informativo del Ministerio de Educación Nacional de Guinea Ecuatorial* 7 (March 8, 1970), n.p.

23. U.S. State Department, *Background Notes,* 1986, p. 1.

24. See República de Guinea Ecuatorial, Ministerio de Planificación y Desarrollo Económico. Dirección General de Estadística. *Informe General de al estadística educativa para la ensenañza del frances.* Report prepared by Luc Cohen. Malabo, July 1986.

25. U.N. Development Program, "Tendencias estrategias y prioridades de cooperación technica." (Planning report for the Third Donor's Conference on Equatorial Guinea, Malabo, 1987), n.p.

26. *Ibid.*

27. Donato Ndongo-Bidyogo, *Antología de la literatura Ecuatoguineana* (Malabo: Editora Nacional, 1984). Also see Constantino Ocha'a Mve Bengobesama, *Semblanzas de la Hispanidad* (Madrid: Ediciones Guineanos, 1985).

28. Guinée Equatoriale, "Le president Obiang Nguema Souhaite une aide intensive de la France," *Marchés Tropicaux* 2238 (Sept. 30, 1988), p. 2609.

29. U.S. Department of State, "Annual Report on Human Rights Practices in Equatorial Guinea" (Washington, D.C.: Government Printing Office, 1986), n.p. I am indebted to Earl Irving, desk officer for Equatorial Guinea for the document.

30. Anacleto Olo Mibuy, "¿Es posible un espacio hispanófono en Africa?" *Africa 2000,* año 3, epoca 2, num. 5, p. 38.

31. Donato Ndongo-Bidyogo, "Hispanidad," *Africa 2000,* año 2, epoca 2, num. 6, p. 3.

32. Program of the Christian Women's Association of Equatorial Guinea, Malabo, May 1988; Trinidad Morgades Besari, president.

33. Klinteberg, *Equatorial Guinea,* p. 6.

34. *Ibid.*

35. Fernandez, "The Affirmation of Things Past," p. 450.

36. *Ibid.,* p. 85.

37. Teodoro Obiang Nguema, *Pensamiento político del Presidente Obiang Nguema Mbasogo por discursos y citas* (Malabo: Departamento de Prensa y Medios de Comunicación Social de la Presidencia del Gobierno, 1982), p. 140.

38. U.N. Development Program, "Tendencias, estrategias y prioridades de cooperación tecnica," n.p.

39. Purificación Angué Ondó, "Cartas de Nuestros Lectores," *Africa 2000,* año 1, epoca 2, num. 2/3, p. 46.

40. C. López Monís, "Aspectos de la lucha sanitaria en Guinea," *Archivos de Instituto de Estudios Africanos* 9 (August 1949), p. 17.

41. Roland Lamusse, "Labour Policy in the Plantation Islands," *World Development* 7, 12 (1980), p. 1036.

42. Manuel de Teran, *Síntesis geográfica de Fernando Póo* (Madrid: Consejo Superior de Investigaciones Cientificas, Instituto de Estudios Africanos, 1962), p. 87.

43. *Ibid.,* pp. 58–61.

44. A. Arbelo Curbelo and R. Villarino Ulloa, *Contribución al estudio de la depoblación indígena en los territorios españoles del Golfo de Guinea, con particularidad en Fernando Póo* (Madrid: Consejo Superior de Investigaciones Cientificas, Instituto de Estudios Africanos, 1942), p. 100.

45. López Monís, "Aspectos," p. 17.

46. J. C. Caldwell, G. E. Harrison, and P. Ouiggan, "The Demography of Micro-States," *World Development* 8, 12 (1980), p. 953.

47. Dan Lantum, "Demographic Transition of Cameroon between 1900 and 1982, with Special Reference to Natality and Mortality" (Paper delivered at the Deuxieme Journées Medicales de Yaoundé, Yaoundé, January 23–28, 1983), p. 7.

48. Balandier, *Sociology*, p. 102.

49. Carlos Krohnert Nchama, "La mortalidad infantil y juvenil en Guinea Ecuatorial," Africa 2000, año 3, epoca 2, num. 5, p. 29.

50. Jacques May, *The Ecology of Malnutrition in Eastern Africa and Four Countries of Western Africa* (New York: Hafner Publishing Company, 1970), p. 34.

6
Conclusion

In Equatorial Guinea the capricious violence of the first decade of independence has been replaced by a quieter authoritarianism. The internal issue of human rights is still subsidiary to the strategic and economic goals of outsiders. France, South Africa, and Nigeria, among others, have differing and conflicting aims that supersede any abstract concern with the conditions of the majority of Guineanos.

The achievements of the Obiang Nguema regime have been significant. The question is, will they be enough? In terms of human rights, the situation has improved. They could hardly have been worse. It is important to remember that "nowhere in modern times has a tyrant [Macias Nguema] . . . managed to destroy his country and annihilate his own people so extensively and persistently in the full knowledge, if not with the assistance of the nation's former colonizer; of its refugee-crowded neighbors, Nigeria, Cameroon, and Gabon; of its allies and/ or aid partners, the U.S.S.R., Cuba, China, France, the U.N. organizations, and the European Economic Community; and of the Vatican, not to mention with the tacit acquiescence of the OAU."[1] Another observer concluded that "at least the possibility of important uranium [and oil] deposits might supply one answer to the intriguing question about the international silence on Equatorial Guinea. . . . After all, oppressive rule in, for instance, Uganda and Chile has been taken up by either the Eastern or Western bloc."[2] Many knew of excesses in Equatorial Guinea but few were interested.

Various organizations continue to cite serious abuses. Amnesty International has condemned the sentencing of civilians to death by military courts. The European Economic Community is now trying to use its economic leverage to assure a transition to a working democracy in the 1990s. At present, Equatorial Guinea is still an example of the triumph of the right of nations over the rights of individuals.

There are two ways of viewing this sad and cautionary past. One is to see the first postindependence government as an aberration. Macias

Nguema can be viewed as a psychotic who pulled the nation off its economic and social trajectory. Alternatively, postindependence governments can be seen as essentially the same, varying only in the intensity of their terror. Indeed, one scholar would say that the chief characteristic of the first twenty years of independence has been continuous "Nguemism."[3]

Both of these views have some validity. The first dictatorship was fueled by extreme paranoia. However, the idiosyncratic elements of the Macias Nguema dictatorship should not blind us to the fact that the political system worked, in its way, for eleven years. The first dictatorship promised something and gave something. If madmen become leaders, the question is why do others follow? In Equatorial Guinea the leader was able to promise rewards for a number of years based only on the plunder of the remnants of the colonial economy. His values, however bizarre, initially were not in conflict with some of the core values of traditional Fang culture. It was only when he traversed the line between the acceptable and the unacceptable within that framework that he was deposed. He crossed that line when he began to kill members of his own terror apparatus.

From the point of view of the men of Mongomo, terror had a function. It kept a check on the centrifugal forces threatening to split the country apart. The state had neither economic nor ethnic unity. No national consensus existed and leaders could not find one. Traditional values and symbols had importance among a large segment of the population but less meaning among minority ethnic groups and members of the educated elite. The first years of independence were an attempt to create a new consensus by returning to the past. The process also involved physically removing groups or persons not in harmony with this movement.

The present government is a continuation of the old regime in terms of personnel but not in terms of the level of terror. Internally it represents a toning down rather than a shift in the use of state violence. Changes have been prompted more by a desire for outside aid than by appeals to conscience.

Some commentators have attributed the economic problems of the country to the dictatorial governments it has endured. The microstate "is an excellent test of the thesis that what Africa needs is a dose of old-fashioned Western capitalism." According to this analysis, "the disincentives [to development] include not only the vagaries of world markets for cocoa, coffee and tropicals hardwoods, but the local hardliners who prefer State trading, fiscal deficits and predatory capitalism."[4] However, this is too simplistic. Are the present problems of the country the result of the postindependence regimes or are they inherent in the structure of the microstate vis-à-vis the world system? Even with all the aid possible, is Equatorial Guinea a viable entity?

The salvation and the curse of Equatorial Guinea is its lack of population. Unlike many other microstates, it is not overcrowded. Underpopulation brings problems of its own. The country lacks a large domestic market. It also lacks the population to make plantation agriculture a real option for the future. From the 1820s to the present, attempts to tie labor to agriculture have resulted in labor abuse, regardless of the nationality or ideology of the users of labor. Any attempt to revive the pre-1968 economic system, however well meant, would run the risk of coercing workers.

The system that made Equatorial Guinea a model colony is no longer practical. A protected metropolitan market no longer exists. The country is at last coming to this realization. The Ministry of Agriculture is trying to encourage diversification. Prospects include sugar, cotton, and sisal, as well as maize, rice, soybeans, and peanuts. Development of such crops would provide foreign exchange as well as lessen dependence on imported food. In 1988 Obiang Nguema announced that his country was "pursuing a plan to diversify agriculture in order to provide the population with consumer goods so that our people do not feel the effects of hunger or suffer shortages of goods which often cause problems in many countries." He also noted that, "if the country managed to achieve total self-sufficiency in these products [e.g., malanga] and produce a surplus, this could be directed towards other national markets, beginning with those of neighbouring countries."[5]

The republic must find further alternatives to plantation agriculture; this will not be easy. The current world price of oil does not make it worth exploiting in Equatorial Guinea. Tourism, which showed some promise twenty years ago, is moribund because of a lack of infrastructure. Although the black sand beaches are attractive, a place like Bioko has little competitive advantage in a world of tropical islands. Furthermore, the presence of a highly suspicious police-state apparatus does not foster an influx of foreign travelers. In the short run, forestry will provide the greatest export earnings. It is not labor intensive and some virgin forest remains in Rio Muni. However, in the long run, unless this resource is replenished, the little country will open itself to financial and ecological disaster.

In the future, the republic's ability to create a raison d'être will be paramount. It is a sovereign state with less population than the Gaza Strip and a land area roughly equal to the U.S. state of Maryland. Malabo has argued that, even with its small size, it can be the intermediary between the African and South American continents. It has been proposed that it convert itself into an African Bimini. Greater investment could be made in communications and in tourism. This would involve the creation of a legal framework that would benefit financiers looking for a tax haven and a relaxed legal code. The government could encourage service activities related to regional commercial transactions (e.g., transit deposits and free-trade zones). It could also adopt special statutes for international banking and provide flag of convenience facilities. To adopt such a course would at some point demand reduced participation in the CFA zone and the Central African Customs Union.[6]

To date Equatorial Guinea bears witness to the durability of colonial frontiers. Respect for those frontiers has had contradictory consequences. Rio Muni's borders divide the Fang and at the same time put some of them in a union with the Bubi located 150 miles away. Consequently, Equatorial Guinea is a small, multiethnic state; ethnic rivalry is liable to play a role in the future. It may produce tension but not fragmentation. Although there are émigré groups that urge Bubi separatism, under-populated Bioko would find it hard to sustain a separate existence. Nepotism and infighting among the ruling Esangui is far more likely.

Although the elite that now rules Equatorial Guinea has an ethnic core, it has co-opted important members of other groups. Politics involves an ethnic balancing act, which makes the situation unstable. If access to power is too narrowly constricted, members of other segments of the dominant ethnic group will attempt to alter the situation. Obiang Nguema has partially dealt with this threat by listening to members of his hometown clique. He has, at the same time, insulated himself from popular upheaval through the use of a Moroccan praetorian guard. Unlike Macias, he appears well aware of the need to balance force with rewards.

In the final analysis, the future of the state may rest on the attitudes of its African neighbors. Up to now the surrounding states have been beset by their own problems of national integration. The immutability of boundaries, a principle enshrined in the OAU charter, has militated against overtly annexationist policies. This may change if the regime's financial needs lead it to certain alliances. An adviser to the Nigerian government has said that "it just may well be that Equatorial Guinea as presently constituted poses a danger not only to herself but to Nigeria and the rest of independent Africa. The Indian government's dismemberment of Bangla Desh from Pakistan may provide a useful lesson for us."[7]

The South African presence in the Bight of Biafra has sent shock waves through certain African governments. It has also created more interest in Equatorial Guinea than human rights abuses ever did. Given the amount and consistency of Nigerian opposition, it is probable that Malabo will try to come to a modus vivendi with its neighbor. This will depend on how much Lagos is willing to grant to the smaller state. Because some agricultural projects depend on Nigerian labor, Malabo may not have much room to maneuver. In a world in which island nations are very vulnerable to quick strikes by national or mercenary forces, it makes little sense for Equatorial Guinea to offend a nation with ample grounds for intervention.

As seen, the need for constant inputs of aid has pulled Equatorial Guinea into dangerous alignments. At present the country is unable to sustain itself without outside patronage. Given this need, true independence is perhaps a fatuous hope. One question remains: Will Equatorial Guinea gradually integrate itself with its African neighbors or will it seek support farther afield?

NOTES

1. René Pélissier, "Autopsy of a Miracle," *Africa Report* 25, 3 (May–June 1980), p. 10.

2. Suzanne Cronjé, *Equatorial Guinea, The Forgotten Dictatorship: Forced Labour and Political Murder in Central Africa* (London: Anti-slavery Society, 1976), pp. 36–37.

3. Max Liniger-Goumaz, *Connaître, la Guinée Equatoriale* (Geneva: Editions des Peuples Noirs, 1986), p. 86.

4. *African Contemporary Record, 1985–1986*, ed. Colin Legum (New York: Africana Publishing Company, a division of Holmes and Meier, 1987), p. B227.

5. *Courier* (Nigeria) 107 (January–February 1988), p. 38.

6. Marcos-Manuel Ndongo, "Guinea Ecuatorial: Posibilidades de desarrollo económico," *Africa 2000*, año 3, epoca 2, num. 6, p. 23.

7. Akinjide Osuntokun, "The Security Aspect of Nigeria–Equatorial Guinea Relations." (Paper delivered at the Nigerian Institute of International Affairs Seminar, April 28, 1987), p. 19.

Selected Bibliography

Adams, John. *Remarks on the Country Extending from Cape Palmas to the River Congo.* London, 1823; reprint, London: Frank Cass and Co., 1966.

African Agricultural Association. *Prospectus,* 1842 (n.p.).

African Contemporary Record. Ed. Colin Legum (London: Rex Collings; New York: Africana Publishing Company, a division of Holmes and Meir, annual).

Akinyemi, Bolaji. "Nigeria and Fernando Poo, 1958–1966: The Politics of Irredentism." *African Affairs* 69, 276 (July 1970), pp. 236–249.

Alianza Nacional de Restauración Democrática (ANRD). *Informational Bulletin* (Geneva: ANRD, 1974).

Allen, William, and Thomas Thompson. *Narrative of the Expedition Sent by Her Majesty's Government to the River Niger, in 1841, Under the Command of Capt. H. D. Trotter.* 2 vols. London, 1848; reprint, New York: Johnson Reprint Corporation, 1967.

Almonte y Muriel, Enrique de. *Someras notas para contribuir a la descripción física geológica y agrícola de la zona noroeste de la isla de Fernando Póo y de la Guinea continental española.* Madrid: Imprenta y Litografía de Deposito de la Guerra, 1902.

Amnesty International. *Equatorial Guinea: Military Trials and the Use of the Death Penalty.* New York: Amnesty International, 1987.

_____. *Amnesty International Report, 1987.* London: Amnesty International Publications, 1988.

Arbelo Curbelo, A., and R. Villarino Ulloa. *Contribución al estudio de la depoblación indígena en los territorios españoles del Golfo de Guinea, con particularidad en Fernando Póo.* Madrid: Consejo Superior de Investigaciones Cientificas, Instituto de Estudios Africanos, 1942.

Arija, Julio. *La Guinea española y sus riquezas.* Madrid: España Colpe, SA, 1930.

Armengol y Cornet, Pedro. *¿A las islas Marianas o al Golfo de Guinea?* Madrid: Imprenta y Libreria de Eduardo Martínez, 1878.

Artucio, Alejandro. *The Trial of Macias in Equatorial Guinea: The Story of a Dictatorship.* Geneva: International Commission of Jurists and International University Exchange Fund, 1980.

155

Aymemí, Antonio. *Los bubis en Fernando Póo, Colección de artículos publicados en la revista colonial la "Guinea española."* Madrid: Galo Saez, 1942.

Balandier, Georges. *The Sociology of Black Africa: Social Dynamics in Central Africa.* London: Andre Deutsch, 1970.

Baumann, Oscar. *Eine afrikanische Tropen-Insel: Fernando Poo und die Bube.* Vienna: E. Hölzel, 1888.

Berman, Sanford. "Spanish Guinea: Enclave Empire." *Phylon* 17 (December 1956), pp. 349–364.

Boneke, Juan Balboa. *Sueños en mi Selva, antología poética.* Malabo: Centro Cultural Hispano-Guineano, 1987.

Bravo Carbonell, Juan. *Fernando Póo y el Muni: Sus misterios y riquezas. Su colonización.* Madrid: Imprenta de "Alrededor de Mundo," 1917.

———. "Possibilidades económicas de la Guinea española." *Boletín de la Sociedad Geográfica Nacional* 72, 8 (August 1933), p. 525.

———. *Territorios españoles del Golfo de Guinea.* Madrid: Imprenta Zoila Ascasibar, 1929.

Brooks, George. *The Kru Mariner in the Nineteenth Century.* Newark, Del.: Liberian Studies Association, 1972.

Brown, Robert. "Fernando Po and the Anti-Sierra Leonean Campaign, 1826–1834." *International Journal of African Historical Studies* 6, 2 (1973), pp. 249–264.

Burton, Sir Richard. *Abeokuta and the Cameroons.* 2 vols. London: W. Cloves and Sons, 1863.

———. *A Mission to Gelele, King of Dahome.* London: Tinsley Brothers, 1864.

———. *Two Trips to Gorilla Land and the Cataracts of the Congo.* 2 vols. London: Low and Searle, 1876.

Buxton, Sir Thomas Fowell. *The African Slave Trade and its Remedy.* London, 1840; reprint, London: Dawsons of Pall Mall, 1968.

———. *Memoirs.* Ed. Charles Buxton. London: John Murray, 1848.

Caldwell, J. C., G. E. Harrison, and P. Ouiggan. "The Demography of Micro-States." *World Development* 8, 12 (1980), pp. 953–967.

Cencillo de Pineda, Manuel. *El brigadier Conde de Argelejos y su expedición militar á Fernando Póo en 1778.* Madrid: Consejo Superior de Investigaciones Cientificas, Instituto de Estudios Africanos, 1948.

Chamberlin, Christopher. "The Migration of the Fang into Central Gabon During the Nineteenth Century: A New Interpretation." *International Journal of African Historical Studies* 11, 3 (1978), pp. 429–456.

Church, R. J. Harrison et al. *Africa and the Islands.* 4th ed. New York: John Wiley and Sons, Inc., 1977.

Clarence-Smith, Gervase. "The Impact of the Spanish Civil War and the Second World War on Portuguese and Spanish Africa." *Journal of African History* 26 (1985), pp. 309–326.

Coll, Armengol. *Segunda memoria de las misiones de Fernando Póo y sus dependencias.* Madrid: "Imprenta Iberia" de E. Maestre, 1911.

Cordero Torres, José Maria. *Tratado elemental de derecho colonial español.* Madrid: Editora Nacional, 1941.

Cronjé, Suzanne. *Equatorial Guinea, The Forgotten Dictatorship: Forced Labour and Political Murder in Central Africa.* London: Anti-slavery Society, 1976.

Davis, R. W. *Ethnohistorical Studies on the Kru Coast.* Newark, Del.: Liberian Studies Association, 1976.

Decalo, Samuel. *Psychoses of Power.* Boulder, Colo.: Westview Press, 1988.

De los Ríos, Juan Miguel. *Memoria sobre las islas de Fernando Póo y Annobón.* Madrid: Sociedad Económica Matritense, 1844.

De Maret, P. "Fernanado Po and Gabon." In *The Archaeology of Central Africa,* ed. F. Van Noten. Graz, Austria: Akademische Druck-u. Verlagsantalt, 1982.

Dike, Kenneth. *Trade and Politics in the Niger Delta, 1830–1885.* Oxford: Clarendon Press, 1956.

Dirección General de Marruecos y Colonias. *Anuario de estadística y catastro de la dirección de agricultura, 1944.* Madrid: Dirección de Agricultura de los Territorios Españoles del Golfo de Guinea, 1944.

Dommen, Edward. "Some Distinguishing Characteristics of Island States." *World Development* 7, 12 (December 1980), pp. 931–943.

Economist Intelligence Unit. *Congo, Gabon, Equatorial Guinea. Country Report. Analyses of Economic and Political Trends Every Quarter.* London: Economist Intelligence Unit [quarterly].

——— . *Gabon, Equatorial Guinea. Country Profile, Annual Survey of Political and Economic Background.* London: Economist Intelligence Unit [annual].

——— . *Quarterly Economic Review of Gabon, Congo, Cameroon, CAR, Chad, Equatorial Guinea.* London: Economist Intelligence Unit [quarterly].

Fegley, Randall. *Equatorial Guinea: An African Tragedy.* New York: Peter Lang, 1989.

Fernández, Cristobal. *Misiones y misioneros en la Guinea española.* Madrid: Editorial Co. SA, 1962.

Fernandez, James W. "The Affirmation of Things Past: Alar Ayong and Bwiti as Movements of Protest in Central and Northern Gabon." In *Protest and Power in Black Africa,* eds. Robert Rotberg and Ali Mazrui, pp. 427–457. New York: Oxford University Press, 1970.

——— . *Bwiti: An Ethnography of the Religious Imagination in Africa.* Princeton, N.J.: Princeton University Press, 1982.

——— . "Fang Representations Under Acculturation." In *Africa and the West, Intellectual Responses to European Culture,* ed. Philip Curtin, pp. 3–48. Madison: University of Wisconsin Press, 1972.

Fernández, Rafael. *Guinea, Materia Reservada.* Madrid: Sedmay, 1976.

Garcia Domínguez, Ramón. *Guinea, Macías, la ley del silencio.* Barcelona: Plaza y Janes, 1977.

Garcia-Trevijano, Antonio. *Toda la verdad, mi intervención en Guinea.* Barcelona: Ediciones Dronte, 1977.

Gard, Robert. "The Colonization and Decolonization of Equatorial Guinea." Pasadena, Calif., unpublished manuscript, 1974.

——— . "Equatorial Guinea: Machinations in Founding a National Bank." In *Munger Africana Library Notes* 27. Pasadena: California Institute of Technology, October 1974.

Green, Lawrence. *Islands Time Forgot*. London: Putnam, 1962.

Góngora Echenique, Manuel. *Angel Barrera y las posesiónes españolas del Golfo de Guinea*. Madrid: San Bernando, 1923.

González Echegaray, Carlos. *Estudios Guineos*. 2 vols. Madrid: Consejo de Investigaciones, Instituto de Estudios Africanos, 1964.

Guillemard de Aragón, Adolfo. *Opúsculo sobre la colonización de Fernando Póo y revista de los principales establecimientos europeas en la costa occidental de Africa*. Madrid: Imprenta Nacional, 1852.

Hahs, Billy Gene. "Spain and the Scramble for Africa, the 'Africanistas' and the Gulf of Guinea." Ph.D. diss., University of New Mexico, 1980.

International Cocoa Organization. *Quarterly Bulletin of Cocoa Statistics* 4 (September 1985).

Jackson, Robert, and Carl G. Rosberg. *Personal Rule in Black Africa*. Berkeley: University of California Press, 1982.

Jakobeit, Cord. "Aquatorialguinea: Schwierge Rehabilitation." In *Afrika Spectrum*. Hamburg: Institut für Afrika-Kunde, no. 2, 1987.

————. "Equatorial Guinea: Long Term Viability of Cocoa Production on Bioko." Unpublished report for the World Bank Cocoa Rehabilitation Project, Hamburg, 1987.

Johnston, Harry. *George Grenfell and the Congo*. New York: D. Appleton and Company, 1910.

Klinteberg, Robert af. *Equatorial Guinea—Macías Country*. Geneva: International University Exchange Fund Field Study, 1977.

Kobel, Armin. "La République de Guinée Equatorialle, ses ressources potentielles et virtuelles. Possibilités de développement. Ph.D. diss., Université de Neuchatel, 1976.

Krohnert Nchama, Carlos. "La mortalidad infantil y juvenil en Guinea Ecuatorial." *Africa 2000*, año 3, epoca 2, num. 5, pp. 28–33.

Lamusse, Roland. "Labour Policy in the Plantation Islands." *World Development* 7, 12 (1980), pp. 1035–1050.

Liniger-Goumaz, Max. *Connaître la Guinée Equatoriale*. Geneva: Editions des Peuples Noirs, 1986.

————. *Guinée Equatoriale, de la dictature des colons à la dictature des Colonels*. Geneva: Editions de Temps, 1982.

————. *Historical Dictionary of Equatorial Guinea*. Metuchen, N.J.: Scarecrow Press, 1979.

————. *ONU et dictatures. De la démocratie et de droits de l'homme*. Paris: Editions L'Harmattan, 1984.

————. *Statistics of Nguemist Equatorial Guinea*. Geneva: Editions du Temps, 1986.

————. *Small Is Not Always Beautiful: The Story of Equatorial Guinea*, trans. John Wood. London: C. Hurst and Company, 1988.

López Monís, C. "Aspectos de la lucha sanitaria en Guinea." *Archivos de Instituto de Estudios Africanos* 9 (August 1949), pp. 7–16.

López Perea, Enrique. *Estado actual de los territorios españoles de Guinea*. San Fernando, Spain: J. M. Gay, 1905.

Lo que es y lo que prodrá ser la Guinea española. Barcelona: "El Misionero," 1931.
Magaz, Manuel Fernández. *Cuentos en el Abáa.* Malabo: Centro Cultural Hispano-Guineano, 1987.
Martín del Molino, Amador. "Que sabemos actualmente del pasado de Fernando Póo." *La Guinea española* (March 15, 1962), pp. 67–68.
May, Jacques. *The Ecology of Malnutrition in Eastern Africa and Four Countries of Western Africa.* New York: Hafner Publishing Company, 1970.
Nvó Mbomio, Ponciano. "Relaciones entre la educación y el desarrollo económico." *Organo informativo del Ministerio de Educación Nacional de Guinea Ecuatorial* 7 (March 8, 1970).
Memoria que presenta a las cortes el ministro de estado respecto a la situación política y económica de las posesiones españoles del Africa Occidental en el año de 1915. Madrid: Imprenta de Juan Perez Torres, 1915.
Moreno-Moreno, José Antonio. *Reseña histórica de la presencia de España en el Golfo de Guinea.* Madrid: Consejo Superior de Investigaciones Cientificas, Instituto de Estudios Africanos, 1952.
Mountjoy, Alan B., and Clifford Embleton. *Africa: A New Geographical Survey.* New York: Praeger, 1967.
Ndongo-Bidyogo, Donato. *Historia y tragedia de Guinea Ecuatorial.* Madrid: Editorial Cambio 16, 1977.
_____. "Lengua e identidad." *Africa 2000,* año 2, epoca 2, num. 2/3, p. 3.
_____. *Antología de la literatura Ecuatoguineana.* Madrid: Editora Nacional, 1984.
Ndongo, Marcos-Manuel. "Guinea Ecuatorial, Posibilidades de desarrollo económico." *Africa 2000,* año 3, epoca 2, num. 6, pp. 16–25.
Nosti Nava, Jaime. *La agricultura en Guinea española.* Madrid: Dirección General de Marruecos y Colonias, 1955.
Nze Abuy, Rafael M. *Familia y matrimonio fan.* Madrid: Ediciones Guinea, 1985.
Obiang Nguema, Teodoro. *Pensamiento político del Presidente Obiang Nguema Mbasogo por discursos y citas.* Malabo: Departmento de Prensa y Medios de Communicación Social de la Presidencia del Gobierno, 1982.
_____. *Guinea Ecuatorial, País Joven.* Malabo: Ediciones Guinea, 1985.
Ocha'a Mve Bengobesama, Constantino. *Guinea Ecuatorial, polémica y realidad.* Madrid: Ediciones Guinea, 1985.
_____. *Semblanzas de la Hispanidad.* Madrid: Ediciones Guinea, 1985.
Ochaga Ngomo, Buenaventura. "Nacimiento de la libertad de Guinea Ecuatorial." *Organo informativo del Ministerio de Educación Nacional de Guinea Ecuatorial* 7 (March 8, 1970).
Olo Mibuy, Anacleto. "¿Es posible un espacio hispanófono en Africa?" *Africa 2000,* año 3, epoca 2, num. 5, pp. 33–35.
Osoba, S. O. "The Phenomenon of Labour Migrations in the Era of British Colonial Rule: A Neglected Aspect of Nigeria's Social History." *Journal of the Historical Society of Nigeria* 4, 4 (1969), pp. 515–538.
Osuntokun, Akinjide. *Equatorial Guinea–Nigerian Relations.* Ibadan: Oxford University Press for the Nigerian Institute of International Affairs, 1978.
_____. "Nigeria–Fernando Po Relations from the Nineteenth Century to the Present." Paper delivered at the Canadian African Studies Association Conference, Université de Sherbrooke, Quebec, April 26–May 3, 1977.

———. *Nigeria in the First World War.* Atlantic Highlands, N.J.: Humanities Press, 1979.

———. "The Security Aspect of Nigeria–Equatorial Guinea Relations." Paper delivered at the Nigerian Institute of International Affairs, Lago, April 28, 1987.

Pélissier, René. "Autopsy of a Miracle." *Africa Report* 25, 3 (May–June 1980), pp. 10–14.

———. "Fernando Póo: Un archipel hispanoguineen." *Revue française d'études politiques africaines* 33 (September 1968), pp. 80–112.

———. "Spanish Guinea: An Introduction." *Race* 6, 2 (1964), pp. 117–128.

———. "Spain's Discreet Decolonization." *Foreign Affairs* 43, 3 (April 1965), pp. 519–527.

———. "Uncertainty in Spanish Guinea." *Africa Report* 13, 3 (March 1968), pp. 16–38 *passim.*

Pereira Rodriguez, Teresa. "Las relaciones maritimo-comerciales entre Canarias y los territorios del Golfo de Guinea (1858–1930)." In *Las Canarias y Africa (Altibajos de una gravitación),* ed. Victor Morales Lezcano, pp. 51–77. Las Palmas de Gran Canaria: Cabildos de Gran Canaria, 1985.

———. "Notas sobre el colonialismo español en el Golfo de Guinea (1880–1912)." *Estudios Africanos, Revista de la Asociatión Española de Africanistas* 1, 2d semester (1985), pp. 92–107.

Perpiña Grau, Romám. *De colonización y economía en la Guinea Española.* Madrid: Editorial Labor, SA, 1945.

Ramos-Izqierdo, Luis. *Descripción geográfica, y gobierno, administración y colonización de las colonias españolas del Golfo de Guinea.* Madrid: F. Peña Cruz, 1912.

República de Guinea Ecuatorial. Presidencia. *Reseña estadística de la República de Guinea Ecuatorial.* Malabo: Secretaria de Estado para el Plan de Desarrollo Económico y Cooperación, Dirección Tecnica de Estadistica, 1981.

Rio Joan, Francisco del. *Africa Occental española (Sahara and Guinea).* Madrid: Imprenta de la Revista Ténica de Infantería y Caballeria, 1915.

Saavedra y Magdalena, Diego. *Memoria presenta al excelentísimo Senor Ministro de Estado, D. Manuel Allendesalazar.* Madrid: Vincent Rico, 1907.

Silveira, Luis. *Descripción de la isla de Fernando Póo en vísperas del tratado de San Ildefonso.* Madrid: Consejo Superior de Investigationes Cientificas, Instituto de Estudios Africanos, 1959.

Simmons, John, ed. *Cocoa Production, Economic and Botanical Perspectives.* New York: Praeger, 1976.

Sindicato de Promoción de Negocios Industriales y Financieros (de) Madrid. *Fernando Póo y la Guinea continental española.* Madrid: Sindicato de Promoción de Negocios Industriales y Financieros (de) Madrid, 1915.

Sociedad Fundadora de la Compañia Española de Colonización. *Memoria demostrativa de las ventajas y beneficios obtenibles de los territorios españoles del Golfo de Guinea.* Madrid: Imprenta de Fortanet, 1905.

Sundiata, I. K. "Equatorial Guinea: The Structure of Terror in a Small State." In *African Islands and Enclaves,* ed. R. Cohen, pp. 81–100. Beverly Hills, Calif.: Sage Publications, 1983.

———. "Integrative and Disintegrative Terror: The Case of Equatorial Guinea." In *International Terrorism in the Contemporary World*, ed. Marius Livingston, pp. 182–194. Westport, Conn.: Greenwood Press, 1978.

———. "The Roots of African Despotism: The Question of Political Culture." *African Studies Review*, 31, 1 (April 1988), pp. 9–31.

Teran, Manuel de. *Síntesis geográfica de Fernando Póo*. Madrid: Consejo Superior de Investigaciones Cientificas, Instituto de Estudios Africanos, 1962.

Tessmann, Gunter. *Die Bubi auf Fernando Poo: Völkerkundliche Einzelbeschreibung eines westafrikanischen Negerstammes*. Hagen: Filkwan, 1923.

"Treaty Concerning the Conditions of Employment of Nigerian Workers in Fernando Po." *International Labour Review* 38 (August 1943), pp. 238–239.

U.N. Development Program. "Tendencias estrategias y prioridades de cooperación tecnica." Planning report for the Third Donor's Conference on Equatorial Guinea, Malabo, 1987.

Unesco. *Statistical Yearbook, 1986*. Paris: U.N. Educational, Scientific, and Cultural Organization, 1986.

Unzueta, Abelardo de. "Ethnografía de Guinea. Algunos grupos inmigrantes de Fernando Póo." *Africa*, nos. 77–78 (May–June 1948), pp. 28–31.

———. *Geografía histórica de Fernando Póo*. Madrid: Consejo Superior de Investigaciones Cientificas, Instituto de Estudios Africanos, 1947.

———. *Guinea continental española*. Madrid: Instituto de Estudios Políticos, 1944.

———. *Islas del Golfo de Guinea*. Madrid: Instituto de Estudios Africanos, 1945.

Vansina, Jan. "Western Bantu Expansion." *Journal of African History* 25 (1985), pp. 129–145.

Veciana Vilaldach, A. de. *La secta del Bwiti en la Guinea Española*. Madrid: Instituto de Estudios, 1958.

Velarde, Juan, "El plan de desarollo económico y social de Fernando Póo y Rio Muni," *Archivos de Instituto de Estudios Africanos* 71 (1964), p. 14.

Vilar, Juan. "España en Guinea Ecuatorial (1778–1892)." *Anales de la Universidad de Murcia* 28, 3/3 (1969–1970), pp. 265–306.

Volio-Jimenez, Fernando. *Study of the Human Rights Situation in Equatorial Guinea*. Unesco. Commission on Human Rights. 36th sess., February 12, 1980. Item 12 of the provisional agenda, E/Cn.4/1371.

Weinstein, Brian. *Gabon: Nation-Building on the Ogooúe*. Cambridge, Mass.: MIT Press, 1966.

World Bank. *Annual Report, 1987*. Washington, D.C.: World Bank, 1987.

Index

Abá. See Fang, palaver house
Abeso Mondu, Eugenio, 81
Adelfarasin, Hilda, 139
African Christian socialism, 59
African Development Bank, 94, 100, 107, 115, 134
Africa 2000 (magazine), 136
Afriexport, 106
Afri Kara, 56, 125
Afripesca (company), 114
Afro-Portuguese creole. *See* Fa d'Amo
Agency for International Development, 110
Aging, 141–142
Agricultural Chamber of Commerce. *See* Cámara Agrícola
Agriculture, 102, 104, 107, 151. *See also* Export crops; Labor, agricultural; Subsistence farming; *under* Bioko; Bubi; Fang; Rio Muni Province
Ahidjo, Ahmadou, 73, 85
AIDS (acquired immune deficiency syndrome), 144
Air force, 77
Airport. *See under* Bata; Malabo
Akoga wood, 111
Alcohol, 106, 119
ALENA. *See* Compañia Nacional de Colonización Africana
Alianza Nacional de Restauración Democrática (ANRD), 82, 138

Alianza Popular, 83
Alor Ayong movement, 12, 56, 124, 125
Alvarez Garcia, Heriberto, 132
Amin, Idi, 70
Amnesty International, 74, 80, 149
Amphibians, 7
Andeme, Ekong, 73, 74
Angola, 18, 45
Angué Ondó, Purificación, 139
Annobón (island) (Equatorial Guinea), 2, 7, 8, 19, 122(illus.)
  cholera, 68, 144
  city. *See* Palé/San Antonio de Palé
  distance from Malabo, 9
  economy, 9
  fishing boats, 114(illus.)
  flora, 9
  income, per capita, 1
  mountains, 9
  and Portugal, 18
  as provisioning stop, 18
  rainfall, 9
  size and location, 3(fig.), 4(fig.), 9
  and Spain, 18, 19, 28, 30
  and toxic waste, 87
ANRD. *See* Alianza Nacional de Restauración Democrática
Anthropophagy, 130
Apra (company), 107
Arab Aid Funds to Africa, 94
Arab Bank for African Development, 100

Arab Economic Development Bank, 107
Archaeology, 9–10, 13, 14
Association of Equatorial Guinean Women, 138
Ateba, Clemente, 59
Authoritarianism, 2, 40, 87, 88, 149
Axim Consortium Corporation, 87
Ayingono, Oyono, 66
Azamboga/Azapmboga (Cameroon), 11

Babangida, Ibrahim, 86
Baho Chico (Bioko), 49
Bakwiri, 15
Balele (dance), 121
Balenke, 10, 11, 25
Balombe culture, 14
Bananas, 1, 10, 13, 40, 107. See also Plantain
Banco Colonial Español del Golfo de Guinea, 38
Banco de Crédito y Desarrollo (BCD), 105
Banco Exterior de España, 38, 40, 102, 103, 104–105
Bañe, 11
Bañe River, 5
Bank of British West Africa, 38
Bank of Central African States, 99
Banque des Etats d'Afrique Centrale (BEAC), 82, 99, 100, 105
Banque Internationale de l'Afrique Occidental—Guinea Ecuatorial (BIAO-GE), 105
Bantu-speakers, 9, 10, 13, 14, 136
Baptist missionaries, 22, 25, 122
Bapuku, 10, 25
Barerra, Angel, 31, 33
Baseke, 10
Basic Law on the Autonomy of Equatorial Guinea (1963), 57, 58
Bassa, 24, 29
Basupu (Bioko), 35, 60
Basupu del Oeste. See Basupu

Bata (Rio Muni Province), 4(fig.), 6(illus.), 8, 29, 30, 32, 94, 95, 113, 115
airport, 115
art museum, 119
colonial, 34, 35, 40
and France, 34
hospital, 142
military governor, 76
population, 34
port authority, 113
post-independence, 64, 68
prison, 68, 69
schools, 92
settlements, 10
strikes (1966, 1981), 60, 81
water supply, 145
Batanga/Bimbia, 24–25
Batete (Bioko), 49
BCD. See Banco de Crédito y Desarrollo
BEAC. See Banque des Etats d'Afrique Centrale
Beach dwellers. See Playeros
Bebom-Mvet, 124
Beecroft, John, 23
Belgian Congo (now Zaire), 30
Belgium, 101, 115
Benga, 10, 11, 20–21, 25, 29, 120
Benin, Bight of, 23
Benito. See Mbini
Berlin Conference (1884–1885), 29
Betsi, 12, 56
Biafra (Nigeria), 70, 71
War (1967–1970), 71, 97
Biafra, Bight of, 13, 39, 152
islands, 17, 23
BIAO-GE. See Banque Internationale de l'Afrique Occidental—Guinea Ecuatorial
Bidyogo, Ndongo, 80
Biéri (Fang belief system), 124, 129
Bikomo Falls (Rio Muni Province), 95
Bikwele (currency), 99
Bioko (island) (Equatorial Guinea), 1, 2, 3(fig.), 4(fig.), 71, 74, 84

agriculture, 9, 15, 24, 36–38, 41,
 47, 48, 49–50, 97, 107, 108, 109,
 140
cattle ranching, 36, 85, 115
colonial, 1, 17, 18, 20, 21, 25–26,
 28, 32, 35, 36–38, 39, 41, 43,
 44, 46, 47, 48, 49, 51
diseases, 143
economy, 1, 36, 37–38, 39, 45, 51,
 140
ethnic groups, 13–15, 33, 60, 140.
 See also Bubi
exports, 39, 41, 42(table)
geobotanical zones, 9
geology, 8
and Great Britain, 21, 22–23, 24,
 25, 27–28, 45
imports, 39
income, per capita, 1, 140
industry, 1
land, 17, 41, 49, 96, 109
literacy, 1
missionaries, 29
mountains, 8–9
nationalism, 56–57, 58, 60, 61
natural gas, 107
oil, 106
population, 23, 32, 43, 47, 49, 140,
 141, 152
and Portugal, 17, 19
railroad, 35
rainfall, 5
rivers, 35
size, 7–8
and slave trade, 19–20, 21
soil, 9
and Spanish Civil War, 39
wages, 108
and World War II, 40
Birds, 7
Birthrate, 140, 141
Black immigration, 22, 24–25, 26, 36.
 See also Labor, migrant
Boiche, Apolinar, 81
Bokassa, Jean-Bedel, 70
Bolaopi culture, 14
Bomudi, 10

Boneke Balboa, Juan, 136
Bonelli, Emilio, 29
Bongo, Omar, 72, 130
Bongue, 10
Bonkoro, Manuel, 21
Bonkoro I, 21
Bonkoro II, 21
Bonny (Nigeria), 21, 24
Borico, Buale, 81
Bosio Dioco, Edmundo, 62, 63, 67,
 73
 death (1975), 68
Botha, Pieter, 87
Boumba, 10
BRGM. *See* Bureau de Recherches
 Géologiques et Minières
British Baptist Missionary Society,
 22, 25
British West Africa, 43–44, 46
Bubi, 13–15, 131, 152
 agricultural workers, 24
 agriculture, 15, 49, 50
 armed resistance (1898), 33
 assassinations of, 68
 clans, 15
 and European trade, 15, 23
 fishing, 15
 in government, 78
 land, 49
 political organization, 23, 33, 58,
 60
 population, 43
 and Portuguese, 17
 religion, 121, 123, 130
 separatist movement, 60, 62, 82,
 152
 social organization, 15
 and Spanish, 27, 28, 33, 49, 64,
 121
 women, 15
Budget, 94, 99, 102, 103, 133
Buela culture, 14
Buiko, 10
Bujeba, 10, 11
Bulu, 56
Bureaucracy, 66, 99

Bureau de Recherches Géologiques et Minières (BRGM) (France), 114
Bureau of Morocco and Colonies, 31
Bwiti (transethnic African religion), 125–128, 130, 137
    dances, 126, 129
    gods, 126–127
    subgroups, 125
    temples, 125, 126

Caisse Centrale de Coopération Economique (CCE) (France), 104, 115
Cámara Agrícola (Malabo), 31, 37, 38, 41, 107
Cámara Agrícola y Forestal de la Guinea Continental Española (Bata), 35
Cámara de Representantes del Pueblo, 77
Cameroon, 3, 4(fig.), 5, 25, 26, 29, 45, 57, 58, 65, 85, 115, 149
    Central Bank, 75
    cocoa, 42
    and Equatorial Guinea, 72–73, 75, 81, 84, 96, 99, 119
    ethnic groups, 10, 11, 125
    GNP, 98
    labor, 43, 46, 50
    palm oil trade, 24
    volcanic regions, 8
    wages, 108
Canaries (islands), 39, 40, 70
Canoes, 7, 14
    confiscation of, 68, 74, 98
Carbonera culture, 14
Carrero Blanco, Luis, 51, 70
Cassava. See Manioc
Catholic church, 83, 122, 123, 129–130, 131
    missionaries, 18, 28, 29, 121–122, 123, 130. See also Jesuits
Cattle, 18, 85
    ranching, 36, 115
CCE. See Caisse Centrale de Coopération Economique

CEIA. See Comunidad de Españoles con Intereses en Africa
Central bank, 94, 103
Centre International de la Civilisation Bantu (CICIBA), 136
Centro Democrático y Social, 83
CEPSA. See Compañia Española de Petroleos de Guinea Ecuatorial, SA
CFA. See Communauté Financière Africain
Chacón, Carlos, 25, 26
Chamber of the People's Representatives. See Cámara de Representantes del Pueblo
Chantiers Modernes (company), 115
Chevron Petroleum (company), 96
Chile, 149
Cholera, 68, 144
Christian Democratic International, 83
CICIBA. See Centre International de la Civilisation Bantu
Clarence, 21, 23, 25. See also Malabo
Claretian order, 28, 121
Clarian Petroleum (company), 107
Club of Paris, 100, 101
CLUSA. See National Cooperative Business Association
CNLGE. See Cruzada Nacional de Liberación de Guinea Ecuatorial
Coastal ports, 3, 4(fig.), 115
Cocoa, 1, 2, 9, 24, 35, 39, 44, 92, 97, 106, 109, 111(table), 116
    Amelonado, 38
    cooperatives, 49, 110
    diseases, 38
    marketing, 107, 108, 109
    plantations, 24, 36–38, 78, 94, 109
    prices, 40, 41–42
    production, 39, 42, 43, 50, 84, 91, 96, 97, 98, 107, 108, 109(table), 116
    quota system, 36–37
    syndicate, 41
Cocobeach (Gabon), 4(fig.), 5
Coconut palm, 12, 18, 39, 107

Cocotiers (islet), 72
Coffee, 1, 9, 22, 36, 39, 40, 92, 106, 110, 111(table)
  cooperatives, 110
  marketing, 107
  prices, 40, 42
  production, 43, 91, 96–97
  quotas, 42
  syndicate, 41
Colonialism. *See* Spain
Colonial Labor Office. *See* Curadoria Colonial
Combenyamango (Benga chief), 21
Comités de base, 68
Comités de gestion, 110
Communauté Financière Africain (CFA), 84, 99, 100, 104, 105, 108, 115, 152
Communications, 102, 152. *See also* Telecommunications
Comoros Islands, 1, 80, 85
Compagnie Française des Petroles, 106
Compañia Española de Petroleos de Guinea Equatorial, SA (CEPSA), 96
Compañia Forestal de Benito, 35
Compañia Guineana de Navegación Maritima, SA, 115
Compañia Nacional de Colonización Africana (ALENA), 36
Compañia Telefónica Española, 104
Compañia Transátlantica, 29, 36
Comunidad de Españoles con Intereses en Africa (CEIA), 103
Concepción Bay. *See* Riabba
Conceptionist Sisters, 121
Conde de Argelejos, 19
Congo People's Republic, 72, 73
Congüe River, 6
Conjoint secretariat, 62
Conoco. *See* Continental Oil Company
Consejo de la Republica, 64
Constitutional referendum (1968), 62
Continental Oil Company (Conoco), 96

Convergencia Social Democrática (CSD), 82
Cooperatives, 49–50, 110
Coordinating Bureau of Guinean Movements, 57
Copper, 113
  sulphate, 38, 85, 116
Corisco (island) (Equatorial Guinea), 2, 4(fig.), 7, 8, 9
  ethnic group, 20–21
  missionaries, 29
  and Netherlands, 18
  oil, 107
  and Portugal, 18
  settlements, 11
  and slave trade, 20–21, 25
  and Spain, 18, 20–21, 26, 28
Corisco, Bay of, 72
Corvée, 34
Côte d'Ivoire. *See* Ivory Coast
Cotton, 73, 151
Council of Government (1963), 58, 59
  boycott (1966), 60
Council of Ministers, 64
Council of Scientific and Technical Research, 139
Court system, 68, 69
  1963, 58
  1973, 67
  1983, 76–77
Credit, 37, 38, 51, 102, 103, 105, 107, 116
Cruzada de Liberación Nacional, 56
Cruzada Nacional de Liberación de Guinea Ecuatorial (CNLGE), 56
CSD. *See* Convergencia Social Democrática
Cuba, 30, 73, 145, 149
Cuban Press Agency, 70
Curadoria Colonial, 31
Currency, 91, 92, 99, 105, 115–116
"Cycle of the Chronicles of Engong" (Fang epic), 124

Dance, 120–121, 126, 129
Death quintet, 56

Decolonization, 1, 55–63
Defense, Minister of, 76, 79, 82
Deforestation, 112
Delegaciónes de Asuntos Indigenas,
    33, 49
Delmas-Vieljeux (company), 115
Democratic Movement for the
    Liberation of Equatorial Guinea,
    82
Democratic Party of Equatorial
    Guinea (PDGE), 77, 83
Dictatorship. See Macias Nguema,
    Francisco
Dirección General de Marruecos y
    Colonias, 41
Dirección General de Plazas y
    Provincias Africanas, 32, 51
Dirección para la Promocion de la
    Mujer, 139
Disease, 7, 11, 21, 43, 68, 141, 142–
    143, 144
Dissent, 81–83, 88
Drugs, 13, 126
Drums. See Mbeny; Ngom
Duala, 15
Du Chaillu, Paul, 20
Dulu Bon be Afri Kara (Fang myth),
    124–125
Duran Loriga, Juan, 64

Ebang Masie, Eduardo, 77
Ebang Mbele Abang, Antonio, 77
Ebano (Malabo), 119
Ebebiyin (Rio Muni Province), 4(fig.),
    6, 32, 58, 143
Ebolowa (Cameroon), 10
Ebuka, Samuel, 82, 97
Economic and Social Council
    (ECOSOC), 74, 79
ECOSOC. See Economic and Social
    Council
ECU. See European Community Unit
Education, 51, 92, 102, 103, 104,
    131–134. See also under Women
Ekuko (Rio Muni Province), 94
Ela, Norberto, 81
Ela Maye, Florencio, 75, 76, 81

Elections
    1964, 58
    1968, 62–63
    1989, 83
Elf-Aquitaine (company), 106
Elobeys (islands) (Equatorial Guinea),
    2, 9
    plantations, 35–36
    settlements, 10, 11
    and Spain, 29, 30, 35–36
Emancipados, 26, 32–33, 49, 51, 61
Empresa Guineano-Española de
    Petroles, 106
Empresa Nacional Petrolifera, 106
Enahoro, Anthony, 97
Eñeso, Agustin, 56
Ente Autonomo del Porto di Trieste
    (company), 113
Enviko, 10, 25
Environmental study, 119
Equatorial Guinea, Republic of, 1,
    151–152
    area, 1, 2, 3–4(figs.)
    autonomy (1963), 57, 58, 59
    boundaries, 2, 4(fig.), 12, 30, 152
    capital. See Malabo
    colonial (Spanish Guinea), 1, 27,
        30–32, 38–43, 50–52, 57, 58, 59,
        60, 61–62, 131–132. See also
        Decolonization; Religion
    constitution (1968), 62
    constitution (1973), 67
    constitution (1982), 76, 79
    coup (1979), 74–75
    coup attempt (1969), 65, 93
    coup attempt (1981), 78
    coup attempts (1983), 81
    coup attempt (1986), 79, 87
    economy, 2, 41, 43, 77–78, 81, 85,
        91–98, 99–101, 107, 115–116,
        150–151
    income, per capita, 2
    independence (1968), 1, 55, 63
    population, 2, 11, 32, 63, 140, 141,
        151
    president, first. See Macias
        Nguema, Francisco

president, second. *See* Obiang
Nguema Mbasogo, Teodoro
size, 151
state of emergency (1969), 64
and union with Cameroon
proposal, 57, 58
See also Annobón; Bioko; Corisco;
Elobeys; Rio Muni Province
Eria, Apolonio, 131
ESAF. *See* International Monetary
Fund, enhanced structural
adjustment facility
Esangui, 12, 66, 74, 130, 152
Escuela Colonial Indigena, 132
Escuela Nacional de Administración
Publica, 134
Esono, Nguema, 66
Esterias, Cape, 11, 21
Ethnic groups, 79. *See also under*
Bioko; Corisco; Rio Muni
Province
European Community Unit (ECU),
100
European Development Fund, 107,
115, 139
European Economic Community, 75,
94, 100, 101, 106, 112, 114, 115,
149
European trade, 13, 17, 18, 22, 23,
25, 27–28, 29, 106. *See also*
Export crops; Slave trade
Evinayong (Rio Muni Province),
4(fig.), 5, 32, 82
invasion attempt on (1976), 74
Executions, 69, 81, 83, 133
Exigencia (trading company), 78
Export crops, 1, 2, 23–24, 40, 50, 84,
91, 92, 96–97, 98, 99, 139
Exports, 106. *See also* Cocoa; Coffee;
Export crops; Oil; Wood; *under*
Bioko; Rio Muni Province
Eyegue, Miguel, 75

Fa d'Amo (language), 18
Fang (Fang-Fang), 10, 11, 72, 152
agriculture, 12–13, 36, 137
currency, precolonial, 13

dances, 120–121
epics, 124, 127–128
ethnic revivalism, 55, 56
and European trade, 13
fishing, 13
in government, 66, 67, 76, 78, 152
leadership, 11, 66
musical instruments, 120, 124
nationalism, 56, 57, 58, 60, 66
opposition parties, 82
palaver house, 121
population, 11
post-independence, 64
religion, 121, 123–128, 130, 131
social organization, 12
and Spanish, 33–34
tribes, 12
weapons, 13
women, 13, 124–125, 137
See also Alor Ayong movement
Fauna, 7
Federation Internationale des Droits
de l'Homme, 74
Felines, 7
Fernández Magaz, Manuel, 136
"Fernandino," 22, 32
Fernando Po. *See* Bioko
Fincas. See Cocoa, plantations
Fishing, 9, 10, 13, 14, 15, 50, 73, 94,
98, 102, 103, 114
Food and Agriculture Organization,
110, 145
Foodstuffs imports, 106, 107, 151
Foreign aid, 2, 73, 75, 84, 85, 93,
94, 95–96, 100, 101–105, 107,
110, 112, 113, 114, 115, 134, 139,
145, 153
Foreign debt, 99, 100–101, 103, 116
Foreign exchange, 110, 151
Foreign investment, 98, 99–100, 108
Foreign relations, 70–74, 84–87, 94,
152–153. *See also individual
countries*
Forestry, 35, 41, 95, 100, 102, 110–
113, 151
concessions, 112
marketing, 111–112

syndicate, 41
  See also Wood
Forests, 5, 13. *See also under* Rio
  Muni Province
Forsyth, Frederick, 70
Franc (CFA), 99, 105, 108
France, 25, 29, 34, 45, 46–47, 55, 74,
  80, 85, 94–95, 100, 101, 102,
  103–104, 105, 106, 112, 115, 134,
  139, 149
France Cable (company), 104
Franco, Francisco, 2, 30, 31, 39, 40,
  61, 63, 70, 122, 123, 131
Franco–Equatorial Guinean
  Friendship Committee (1986),
  104
FRENAPO. *See* Frente Nacional de
  Liberación de Guinea Ecuatorial
French–Equatorial Guinean Joint
  Commission (1986), 104
Frente de Liberación de Fernando
  Póo, 82
Frente Nacional de Liberación de
  Guinea Ecuatorial (FRENAPO),
  59

Gabon, 3, 5, 6, 9, 29, 65, 115, 149
  colonial, 34
  and Equatorial Guinea, 72, 75, 86–
    87, 99
  ethnic groups, 10–11, 13, 120, 125
  GNP, 98
  labor, 50
  nationalists, 56, 57
  oil, 96, 106
  slave trade, 18, 25
Gabon River, 19
Galvez, José de, 20
Gándara, José de la, 26
Gande/N'Gande (island), 6
Garcia-Trevijano, Antonio, 71, 94
Gay, Alex, 70
GEMSA. *See* Guineana Española de
  Minas, SA
General Assembly (1963), 58, 60
Germany, 28, 29, 45. *See also* West
  Germany

Getty Oil (company), 106
Ghana, 24, 42, 109. *See also* Gold
  Coast
Giscard d'Estaing, Valery, 74
GNP. *See* Gross national product
Gold Coast (now Ghana), 25, 44
Gomes da Silva, Manuel, 17
González, Felipe, 83
Gori Molubela, Enrique, 50, 56, 58,
  59, 61
Gowon, Jakubu, 71
Great Britain, 20, 29, 40, 44, 47, 55
  and Equatorial Guinea, 92, 106,
  107
  See also under Bioko
Great Depression (1930s), 39
Grebo, 24, 44
Gross national product (GNP), 98,
  99, 100, 102
Grupo Macias, 62
Guardia Civil (Spanish), 64
Guardia Colonial, 47
Guillimard de Aragon, Adolfo, 25
Guinea, Gulf of, 4(fig.), 43
Guinea-Bissau, 1
Guineana Española de Minas, SA
  (GEMSA), 114
Guinextebank, 102, 104, 105
Gulf Oil Company of Equatorial
  Guinea, 96

Haile Selassie (emperor of Ethiopia),
  70
Handicrafts, 137, 139
Harp. *See* Ngombi
Health care, 46, 51, 73, 102, 140–145
Hispanicization, 49, 121, 131
Hispanoil (company), 106, 107
Holt, John, and Company, 29
Horn. *See* Nlakh
Human rights, 74, 79–81, 83, 94,
  128, 136, 149
Hydroelectric power, 95, 102, 104,
  115

Iberia (Spanish airline), 95, 103
Ibia, J. Ikege, 29

Iboga, 126, 128
Ibongo, Saturino, 64–65
Idea Popular de Guinea Ecuatorial
    (IPGE), 57, 58, 59, 60, 61
    coalition. See Grupo Macias
Igbo, 71
IMF. See International Monetary Fund
Import duties, 39, 41, 42
Imports, 39, 91–92, 98, 99, 104, 106,
    107, 119
Income, per capita, 50, 51
Industry, 1, 2, 51, 85, 98, 106
Infant mortality, 141, 142
Infertility, 137, 141
INFOGE. See Instituto de Fomento
    de Guinea Ecuatorial
Inprocao (cocoa marketing board),
    108–109
Instituto Cardenal Cisneros (Malabo),
    131–132
Instituto Carlos Lwanga (Bata), 133
Instituto de Fomento de Guinea
    Ecuatorial (INFOGE), 94
Instituto Geológico de Madrid, 103
Integrated Rural Development Project
    (Niefang), 139
Interior, Minister of the, 64, 81
International Agricultural
    Development Fund, 110
International Cocoa Organization,
    108
International Coffee Organization,
    110
International Commission of Jurists,
    74, 79
International Conference of Donors
    for the Economic Reactivation
    and Development of the
    Republic of Equatorial Guinea
    (1982, 1988), 101
International Covenant on Civil and
    Political Rights, 80, 81
    Optional Protocol, 80–81
International Hispanic-African
    Congress (1984), 135
International Labor Organization, 47,
    139

International Monetary Fund (IMF),
    99, 100, 101, 105, 111, 115
    enhanced structural adjustment
    facility (ESAF), 101
International Organization for
    Tropical Woods (1987), 111
International University Exchange
    Fund, 74, 128
IPGE. See Idea Popular de Guinea
    Ecuatorial
Iradier y Bulfy, Manuel, 28–29
Israel, 85, 87
Isubu, 15
Italy, 18, 100, 106, 108, 112, 113,
    115, 145
Itembúe River, 6
Ivanga dances, 120
Ivory Coast, 42, 80, 108

JCFOGE. See Junta Coordinadora de
    las Fuerzas de Oposición de
    Guinea Ecuatorial
Jehovah's Witnesses, 131
Jesuits, 26, 28
Jeune Afrique, 80
John Paul II (pope), 131, 144
Jones, José Luis, 83
Jones, Maximiliano, 37
Jones, Salomé, 51, 56, 59
Juan Carlos (king of Spain), 102
Junta Coordinadora de las Fuerzas de
    Oposición de Guinea Ecuatorial
    (JCFOGE), 82
Juventud en Marcha con Macias, 65,
    66–67, 68, 69, 70, 76

King, Martin Luther, School of
    Administration (Malabo), 134
Kombe, 10, 25
Krohnert, Frieda, 138
Kru, 29, 44

Labor
    agricultural, 24–25, 43–44, 46, 47,
    48, 50, 91, 96, 97–98, 108, 109–
    110, 151
    Asian, 45, 46

colonial, 31, 37, 43–52
compulsory, 98
forced, 46, 79
Japanese, 46
migrant, 1, 34, 43–49, 50, 66, 71,
    84, 91, 97, 108, 109, 153
ordinance (1913), 37
sexual division of, 137
shortage, 107–108, 110
wages, 108, 138
Lack, Hilton, 85
LAGE. See Lineas Aereas de Guinea
    Ecuatorial
Lagos (Nigeria), 23, 25
Land
    colonial policy, 41, 49
    laws, 33, 38–39
    redistribution, 78
    reform, 110
    use, 49
Languages. See Bantu-speakers;
    Benga-speakers; Fa d'Amo;
    French; Kombe-speakers; Spanish
La Salle Brothers missionaries, 132
Law, traditional, 77
League of Nations, 46
Lebanese, 40
Lerena, Juan José de, 25
Liberia, 24, 29, 43, 44–45, 46, 144
Libya, 73
Limbe. See Victoria
Lineas Aereas de Guinea Ecuatorial
    (LAGE), 95
Literacy, 1, 133
Literature, 135, 136
Livestock and Garden Produce
    Project (Bata), 139
Lomé Convention, 108
Lopez, Cape (Gabon), 18, 19, 25
Loreto, Lake. See Moka (lake)
Louis, André, 83
Luba (Bioko), 4(fig.)., 24, 28, 143
    Soviet naval base, 73
Luba (Bubi leader), 33
Lumbering. See Forestry; Wood
Lynslager, William, 23

Macias Nguema, Francisco (Biyogo
    Negue Ndong), 59, 60–61, 62,
    88(n5), 150–151
    arrest (1979), 75
    and economy, 91, 92, 93–94, 96,
        97–98
    and education, 133
    execution (1979), 79, 129
    ideology, 70
    and IPGE, 61
    mayor of Mongomo, 61
    and MONALIGE, 59, 60, 61
    and Munge, 61
    portrait medallion, 67(illus.)
    power base, 66. See also Mongomo
        district
    president (1968–1979), 2, 55, 63–74
    president for life (1972), 67
    and PUNT, 67
    and religion, 128–130
    wives, 138
    youth movement, 64. See also
        Juventud en Marcha con Macias
Mackey, J. L., 29
Macroeconomics, 102
Maho, Luis, 56, 57
Malabo (Bioko, Equatorial Guinea), 7,
    25, 67, 115
    airport, 4(fig.), 85, 95
    anti-independence demonstration
        (1968), 62
    City Council, 32
    colonial, 30, 32, 34, 35, 40
    demonstration (1979), 74
    diseases, 143
    hotels, 85, 119
    lifestyle, 119
    Maternity Hospital, 142
    Nigerians in, 71
    population, 119
    port facilities, 103
    Portuguese in, 70
    post-independence, 64, 85
    prices, 119
    railroad, 35
    restaurants, 119
    schools, 92

Soviet deep sea fishing base, 73
thermal power station, 95
See also Clarence
Malabo (Bubi paramount chief), 33
Malanga, 10, 12, 15, 107, 139, 144, 145, 151
Malaria, 7, 136, 141, 142, 143, 144, 145
Mañe, Acacio, 56
Mangue de Obiang, Constancia, 139
Manioc, 10, 13, 40, 107, 110
Mba, Justino. See Mba Msue, Justino
Mba, Leon, 56
Mba, Moro, 76
Mba Ada, Moses, 64, 78, 94
Mba Msue, Justino, 56, 78
Mba Nchama, Felix, 81
Mbanie (islets), 72
Mba Oñana Nchama, Fructuoso, 81–82, 87
Mba Ovono, Jesus, 59
Mbeny, 120
Mbini (Rio Muni Province), 4(fig.), 34, 35, 64, 95
Mbini River, 3, 4(fig.), 5, 6, 7, 10, 19, 29
Mbomio, C., 74
Mbomio, Leandro, 119
Medical facilities, 85, 142, 143–144
Meke, 12
Methodist missionaries, 29, 122, 131
Metsogo people (Gabon), 125
Mez-m Ngueme, Francisco. See Macias Nguema, Francisco
Micha, Gregorio, 81
Micromesang (Rio Muni Province), 121
Migrations, 10–12, 14, 34, 65. See also Labor, migrant
Miko, Venancio, 81
Military, 76, 77, 85. See also Militia; National Guard
Militia, 68
Minerals, 3, 5, 91, 103, 113, 114
Mining, 102, 113–114
Ministry of Planning and Economic Development, 101–102

Missionaries, 18, 22, 25, 28, 29, 121–122, 124, 130, 132
Mitémele River, 6, 10
Mitterrand, François, 104
MNU. See Movimiento Nacional de Unión
Mobutu Sese Seko, 70, 72
Moka (Bioko), 49, 85, 123
Moka (Bubi paramount chief), 23, 28, 33
Moka (lake), 9
MONALIGE. See Movimiento Nacional de Liberación de Guinea Ecuatorial
Mongomo district (Rio Muni Province), 12, 58, 61, 66, 67, 74, 76, 79, 87, 129, 130, 150
population, 66
Montes de Cristal region, 5, 11
Moran, Fernando, 81
Morgades Besari, Trinidad, 139
Moroccan bodyguard, 77, 81, 87, 119, 129, 152
Morocco, 25, 30, 39, 87
Mortality rate, 51, 65–66, 140, 141, 142, 143
Moto Nsa, Severo, 82, 83
Mountains, 5, 8
Movimiento de Unión de Guinea Ecuatorial (MUNGE), 58–59, 60, 61, 62, 63, 78
coalition. See Grupo Macias
Movimiento de Unión Popular de Liberación de la Guinea Ecuatorial (MUPGE), 57, 58
Movimiento Nacional de Liberación de Guinea Ecuatorial (MONALIGE), 56, 57, 58, 59, 60, 61, 62, 63
coalition. See Grupo Macias
Movimiento Nacional de Unión (MNU), 57, 58
Movimiento Pro-Independencia de Guinea Ecuatorial (MPIGE), 57
Mozambique, 45

MPIGE. *See* Movimiento Pro-
Independencia de Guinea
Ecuatorial
Mpongwe, 10, 11, 120
Muhammad, Murtala, 71
Munga (Benga leader), 21
MUNGE. *See* Movimiento de Unión
de Guinea Ecuatorial
MUPGE. *See* Movimiento de Unión
Popular de Liberación de la
Guinea Ecuatorial
Museums, 119–120
Music, 120, 121, 123–124, 129
Mustrich, Domingo, 25
Mvae, 12
Mvet, 120, 124
Mvoe Ening movement (Bwiti), 130

National Assembly, 62, 76
National Bank of Equatorial Guinea,
92, 95, 105
National Center for Teacher Training,
134
National Committee of Prevention of
AIDS, 144
National Cooperative Business
Association (CLUSA), 110
National Forestry Institute, 95
National Guard, 65, 66, 68, 74
Nationalism, 52, 56–60, 62, 125
National Liberation Crusade. *See*
Cruzada de Liberación Nacional
National Social and Economic
Council, 76
Native Labor Code (1906), 44
Native Trusteeship. *See* Patronato de
Indígenas
NATO. *See* North Atlantic Treaty
Organization
Navy, 76, 77, 85
Ndongo, Buendy, 138
Ndongo, Manuel Ruben, 82
Ndongo, Marcos-Manuel, 105
Ndongo-Bidyogo, Donato, 136
Ndongo Miyone, Atanasio, 56, 57,
59, 60, 61, 62, 63
death, 65

foreign minister, 63, 64–65
Ndowe, 10–11, 58, 127, 130
Necrophagia, 124
Neñe Eñeso, 133
Netherlands, 18, 101, 106
New Calabar (Nigeria), 24
Newspapers, 119
Ngom, 120
Ngoma Nandango, Frederico, 56, 58
Ngombi, 120
Ngomo Nvono, Jesus, 79
Ngoubi, Marien, 70, 72
Nguema, Ela, 66
Nicolls, Edward, 21
Niefang (Rio Muni Province), 4(fig.),
35, 121, 144
Niger Expedition (1841), 22
Nigeria, 1, 3(fig.), 25, 29, 57, 65
cocoa, 42
and Equatorial Guinea, 70–71, 84,
85, 86, 97, 108, 114, 149, 152,
153
labor, 43–44, 47–48, 50
oil, 84, 86, 96
palm oil, 24
population density, 43
See also Biafra, War
Niger River, 22, 23
Nlakh, 120
North Atlantic Treaty Organization
(NATO), 70, 84
Nse Nfumu, Agustin, 139
Nseng, Ela, 76
Nseng, Jaime, 57
Nsomo, Ebengeng, 79
Nsue, Esteban, 56
Nsue, José, 56
Nsue Ela, Pedro, 130
Ntagan ye Nsutmot (Fang Whiteman-
Blackman stories), 127–128
Ntem River, 6, 7, 10, 11, 19
Ntumu, 12
Ntutumu, Eyegue, 66
Ntutumu, Masie, 64, 66, 74
Nuñez de Prado, Miguel, 33
Nvo, Enrique, 56
Nwachukwu, Ike, 86

Nzue Abuy, Rafael, 129, 131

OAU. *See* Organization of African Unity
Obama Eyang, Pablo, 81
Obama Mbain (Fang chief), 33
Obama Owono, Jaime, 81
Obiang Nguema Mbasogo, Teodoro, 2, 66, 74, 75, 78(illus.), 134–135, 149, 150
 bodyguard, 77, 81, 87, 152
 and economy, 98, 101, 102, 103, 104, 114, 116, 151
 opposition to, 81–83, 88
 political philosophy, 77
 president (1979–   ), 75–88
 as pro-Western, 84
 and religion, 130
 second term (1989), 83
 and status of women, 138
 wife, 139
OCGE. *See* Oficina de Cooperación con Guinea Ecuatorial
Oficina de Cooperación con Guinea Ecuatorial (OCGE), 105
Ogowe River, 11
Oil, 3, 70, 104, 106–107, 149
 exploration, 96, 106
 offshore, 1, 72, 106, 107
 prices, 84, 151
Ojukwu, Chukwemeka, 70, 71
Okak, 12
Okume tree, 5, 7, 13, 40, 50, 111
Old Calabar (Nigeria), 21, 24
Olo Mibuy, Anacleto, 136
Onassis-Roussel group, 112
Ondo Edu, Bonifacio, 57, 58, 59, 60, 62, 63, 65
 death (1968), 68
 wife, 138
Ondo Mañe, Damian, 82
Ondo Nguema, Armengol, 79
Oprocage (company), 85
Opus Dei, 94
Organization of African Unity (OAU), 64–65, 71, 72, 73, 74, 82, 93, 149

charter, 152
Council of Ministers, 86
Organization of Petroleum Producing Countries, 107
Osa, Clara, 138
Ossorio y Zayala, Amado, 29
Owen, William, 21
Owono Ndongo, Carmelo, 81
Oyono, Daniel, 82
Oyono, Feliciano, 66
Oyono, Jesus Alfonso, 71, 93

Pahouin, 11, 56, 124, 125
Palagú. *See* Annobón
Palé/San Antonio de Palé (Annobón), 4(fig.), 9
Palm oil, 1, 12, 14, 15, 22, 23–24, 36, 40, 107
 cooperatives, 49
Parcelista. See Tenant farmer
Partido Democrata Popular, 83
Partido Unico Nacional. *See* Partido Unico Nacional de Trabajadores
Partido Unico Nacional de Trabajadores (PUNT), 67, 68
 Women's Revolutionary Section, 68, 138
 youth wing, 68
Patronato de Indígenas, 31, 33
 abolished (1959), 50
PDGE. *See* Democratic Party of Equatorial Guinea
Peanuts, 12, 110, 151
Pechaud et Compagnie Internationale, 114
Pechaud Guinée Equatoriale (company), 114
Penda Recu, Malgalena, 138
People's National Assembly, 67
People's Republic of China, 73, 74, 84, 95, 96, 100, 102, 106, 111, 149
Perea Epota, José, 57
Peres, Shimon, 85
Pico de Basilé, 8
Piedra de Nzas, 5

Plantation economy, 17, 21, 35–38, 41, 43, 49, 66, 151
decline, 91, 97, 116
Playeros, 10
Political parties (1964), 58–59, 62, 77, 79, 82–83
abolished (1970), 67
Political prisoners, 69, 80, 83
Polygyny, 33, 124
Portugal, 17, 18, 55, 70, 104
Postcolonial decline, 2, 91–92, 93, 115–116
PPGE. *See* Progress Party of Equatorial Guinea
Presbyterian missionaries, 29
Presidency (1983 constitution), 76, 77
Presidential palace (Ekuko), 94
Primo de Rivera, Joaquín, 19
Primo de Rivera, Miguel, 31
Príncipe, 1, 3(fig.), 8, 14, 17, 44, 115
GNP, 98
and Spain, 18, 19
Privatization, 115
Progress Party of Equatorial Guinea (PPGE), 82, 83
PROGUINEA (coffee syndicate), 41, 50
Promoport Guinea (Bata port authority), 113
Property code (1948), 49
Prostitution, 137, 139
Protestant missionaries, 122, 124
Public sector, 61, 74
dissidents, 81–82
investment, 94, 102
jobs, 50–51
strike (1966), 60
PUNT. *See* Partido Unico Nacional de Trabajadores
Pygmies, 10

Racism, 51
Railroads, 4(fig.), 34, 35, 36
Ramos de Esquivel, Luis, 17
Red Cross, 71
Religion, 121–131
Reptiles, 7

Riaba (Bioko), 4(fig.), 23, 28, 115
Rio Benito. *See* Mbini River
Rio Campo. *See* Ntem River
Rio de Oro, 39
Rio Muni, 5–6, 9, 10
Rio Muni Province (Equatorial Guinea), 1, 2–3, 4(fig.), 5–7, 29, 74
agriculture, 10, 12–13, 31, 35, 36, 43, 50
calories, 144
capes, 3
capital. *See* Bata
colonial, 10, 25, 27, 30, 31–32, 33–34, 35, 39, 40, 41, 50–51, 56, 60, 122, 125
economy, 1, 35, 50, 51
ethnic groups, 9–13, 25, 33, 34, 55, 58. *See also* Fang
and European trade, 29
exports, 39, 40, 42(table), 50
fauna, 7, 13
fishing, 10, 50
forests, 5, 35, 36, 40, 151
geology, 3, 5
infant mortality, 142
mining, 113–114
nationalism, 56, 57–58, 60
oil, 106
politics, 12
population, 32, 43, 50
Presbyterian churches, 122
Provincial Deputation, 58
rainfall, 5
roads, 35, 36
size, 3
and Spanish Civil War, 39–40
temperature, 5
topography, 3, 5–6
watercourses, 5–7
water supply, 142, 145
Riquelme, Fernando, 105
Rivers, 5–7, 9
Roads, (fig.), 35, 73, 95, 112, 113
Rovira, José, 78
Rubber, 13, 29, 39
Ruiz, Faustino, 32

San Carlos. *See* Luba
Sanitation, 102, 145
San Juan, Cape, 3, 5, 10, 11, 19, 21, 29, 36
Santa Isabel. *See* Malabo (Bioko, Equatorial Guinea)
Santiago de Baney (Bioko), 57
São Tomé, 1, 3(fig.), 8, 17, 24, 36, 38, 44, 115
    GNP, 98
    and Portugal, 18
    and Spain, 18, 19
Sas Ebuera (Bubi chief), 33
Schools, 85, 92, 133, 134, 135
Security apparatus, 69, 79
Semge (company), 112
Seminaries, 121, 122
    seminarians strike (1952), 56
Senyi (Rio Muni Province), 6
Seriche Bioco, S., 78
Seychelles, 1
Shagari, Shehu, 84
Shell Guinea Equatorial (company), 96
Sialo, J. M., 121
Siem (company), 112
Sierra Leone, 22, 24, 25, 44, 45
Silviculture, 112
SIMED, SA, 94
Sindicato Agrícola de Guinea. *See* Sindicato Agrícola de los Territorios Españoles del Guinea
Sindicato Agrícola de los Territorios Españoles del Guinea, 38, 46
Sindicato Maderero, 41
Slaves, 14–15, 18, 22
Slave trade, 10, 17, 18, 19–22, 25
    anti-, 20, 21, 22, 25
Sleeping sickness, 7, 46
Smallpox, 43, 141
Sociedad Colonizadora de Guinea (SOCOQUI), 36
Sociedad Económica de Barcelona, 25
Sociedad Económica Matritense, 25
Sociedad Fundadora de la Compañia Española de Colonización, 30

Societé Belge de Consignation et d'Affrètement Saint Nicolas, 115
Societé Forestière du Rio Muni (company), 95
Societé Française des Dragages et de Travaux Publiques, 95
Societé Planteurs, SA (company), 107
SOCOQUI. *See* Sociedad Colonizadora de Guinea
Sodis-Guinea (company), 107
Sogec (company), 107
Sole National Workers' Party. *See* Partido Unico Nacional de Trabajadores
South Africa, 85–86, 87, 119, 149, 152–153
South Korea, 111
Soviet Union, 73, 84, 98, 100, 101, 106, 119, 149
Spain, 2, 19, 25, 30, 40–41, 44, 55, 64, 100
    and Equatorial Guinea, Republic of, 70, 71, 74, 75, 77, 81, 83, 92, 93–94, 95–96, 100, 101, 102–105, 106, 107, 112, 114, 115, 119, 134, 135, 136
    and NATO, 84
    See also Bioko, colonial; Decolonization; Equatorial Guinea, Republic of, colonial; Rio Muni Province, colonial; Slave trade
Spanish Guinea. *See* Equatorial Guinea, Republic of, colonial
Spanish-Guinean Cultural Center (Bioko), 135, 136
Spanish Institute of Cooperation, 103
Spanish Institute of Foreign Money, 92
Spanish Possessions in West Africa Consultative Council, 30
Spanish Technical Cooperation, 134
Spearheads as currency, 13
Stanley, Henry, 29
Subsistence farming, 1, 107, 110, 137, 145
Supreme Military Council, 76

Supreme Native Tribunal, 33
Syndical Committee for Cocoa, 41
Syndicates, 41
Syphilis, 141

Taiwan, 111
Tamayo, Marcial, 93
Tardiba (company), 95
Taxes, 94, 99, 108
Teachers, 133, 134
Teacher-training institute (Malabo),
    131
Technical Aid Corps (Nigeria), 85
Tecnica Española, 119
Telecommunications, 96, 100, 104
Tenant farmer, 109–110
Territorial Union of Cooperatives, 50
Territories of the Gulf of Guinea, 30
Terror, 2, 66, 68–70, 73, 79, 91, 94,
    138, 149, 150
Three-Year Economic Development
    Plan (1963), 51
Timber. See Wood
Tobacco, 12, 91, 106, 107
Togo, 144
Tonton Macoute (Haiti), 66
Torao, Pastor, 57
Torres Quevado, SA, 96
Tourism, 151, 152
Toxic waste, 87
Trading companies, 36, 40, 78
Transmeditarránea Company, 95, 115
Transportation, 51, 93, 95, 102, 107,
    115, 119
Transport vehicles, 92, 100, 106
Trillos, Antonio (company), 29
Trypanosomiasis, 43, 141, 142, 143
Typhoid fever, 143

UDEAC. See Union Douanière et
    Economique de l'Afrique
    Centrale
Underpopulation, 2, 151
UNDP. See United Nations
    Educational, Scientific, and
    Cultural Organization
Unión Bubi, 58, 62, 123

Unión de Agricultores de la Guinea
    Española, 37
Unión Democrática Fernandina, 58
Union des Populations du Cameroun,
    59
Union Douanière et Economique de
    l'Afrique Centrale (UDEAC), 99,
    100, 101, 105, 152
United Nations, 57, 60, 62, 64, 65,
    72, 74, 80, 81, 86, 96, 100, 108,
    130, 143, 149
    Resolution 2067 (1965), 60
    secretary-general, 74, 93
United Nations Commission on
    Human Rights, 74, 76, 79, 80,
    81, 82, 138
United Nations Committee of
    Twenty-four on Decolonization,
    60, 61
United Nations Development Fund,
    145
United Nations Development
    Program (UNDP), 70, 91, 95,
    100, 101, 102, 112, 114, 133,
    134, 139
United Nations Educational,
    Scientific, and Cultural
    Organization (Unesco),
    conference (1978), 17
United States, 65, 99, 100, 101, 104,
    110, 115, 136
    embassy, 94
    missionaries, 29, 124
    naval mission, 78(illus.)
    oil, 84
Uranium, 5, 103, 113, 149
Ureka region (Bioko), 9, 14
Utamboni. See Mitémele River
Utonde River, 6, 7, 10
Uvomi. See Punta Mbonda

Vai, 44
Varela y Ulloa, José, 19
Vatican, 149
Venereal disease, 21, 46
Victoria (Cameroon), 26
Villages, 11–12, 13, 51, 60, 68, 69

Vivour, William Allen, 24
Volio-Jimenez, Fernando, 80

Water supply, 142, 145
West African Company, 22
West Germany, 86, 99, 101, 106
WHO. *See* World Health Organization
Witchcraft, 127, 130
Woermann Line (company), 28, 29
Women
  adornment, 137
  aging, 141–142
  aid projects, 139–140
  associations, 138–139
  and Bwiti, 137–138
  cooperatives, 110, 139
  dancers, 120, 121
  and education, 136
  health, 136
  in government, 138, 139
  labor, 46, 136–137, 138, 139
  and political parties, 68, 138
  status, 137, 138, 139–140

wages, 138
  See also under Bubi; Fang
Wood, 1, 5, 7, 13, 22, 35, 36, 39, 40,
  42, 50, 92, 94, 95, 106, 111–113,
  151
  production, 110–111, 116
  syndicates, 41
World Bank, 84, 100, 101, 102, 107,
  116, 134, 145
  Cocoa Rehabilitation Project, 78
World Health Organization (WHO),
  93, 144
World Labor Conference, 75
World War I (1914–1918), 45
World War II (1939–1945), 40

Xylophone. *See Mendjang*

Yellow fever, 7, 26, 43, 141
Yriarte, Bernardo de, 20

Zaire, 72, 98
Zither, *See Mvet*

Composed of the mainland province of Rio Muni and the island of Bioko (formerly Fernando Po), Equatorial Guinea is one of the smallest African states and among the newest postcolonial nations. Not until the twentieth century did a European colonial power gain control of either territory. While Spain converted Bioko to a monocultural economy dependent on cocoa, Rio Muni remained largely outside the colonial economy. Franco statist policies guaranteed Bioko planters a degree of affluence in the period after World War II—a prosperity shared by the Bioko cocoa cooperatives but not by the peoples of the mainland. Therefore, when the country achieved independence in 1968, economic development was skewed in favor of the island.

The first president, Francisco Macias Nguema, attempted to break with the economic structures inherited from colonialism, but the increasingly erratic nature of his regime discouraged foreign investment. Moreover, the president-for-life ruled a state in which terror killed many, forced still more to flee, and interfered with the recruitment of Nigerian migrant workers. By the late 1970s, the economy's core—the plantation sector—was in shambles.

In August 1979 the government was overthrown by Macias Nguema's nephew, Colonel Teodoro Obiang Nguema Mbasogo. The new regime changed the formerly anti-Western orientation of the government and encouraged an infusion of foreign capital. Yet despite international aid, the economy remains depressed. The lack of labor and a net loss of population during the Macias Nguema years have contributed to continued economic dislocation. Future recovery is tied to the revival of cocoa and the diversification of the economy.